Monarchs *and* Milkweed

A MIGRATING BUTTERFLY,
A POISONOUS PLANT,
AND THEIR REMARKABLE STORY
OF COEVOLUTION

Anurag Agrawal

PRINCETON UNIVERSITY PRESS
PRINCETON AND OXFORD

Published by Princeton University Press, 41 William Street, Princeton, New Jersey 08540

In the United Kingdom: Princeton University Press, 6 Oxford Street, Woodstock, Oxfordshire OX20 1TR

press.princeton.edu

Jacket photos by Ellen Woods

Library of Congress Cataloging-in-Publication Data

Names: Agrawal, Anurag A.

Title: Monarchs and milkweed : a migrating butterfly, a poisonous plant, and their remarkable story of coevolution / Anurag Agrawal.

Description: Princeton : Princeton University Press, [2017] | Includes bibliographical references and index.

Identifiers: LCCN 2016034053 | ISBN 9780691166353 (hardback : alk. paper)

Subjects: LCSH: Monarch butterfly. | Milkweed butterflies. | Milkweeds. | Coevolution.

Classification: LCC QL561.D3 A47 2017 | DDC 595.78/9—dc23 LC record available at https://lccn.loc.gov/2016034053

British Library Cataloging-in-Publication Data is available

This book was supported in part by Cornell University's Atkinson Center for a Sustainable Future (www.acsf.cornell.edu)

This book has been composed in Perpetua Std

Printed on acid-free paper. ∞

Printed in China

10 9 8 7 6 5 4 3 2 1

CONTENTS

ILLUSTRATIONS

Monarchs *and* Milkweed

CHAPTER 1

Welcome to the Monarchy

You who go through the day
like a wingèd tiger
burning as you fly
tell me what supernatural life
is painted on your wings
so that after this life
I may see you in my night

—Homero Aridjis, "To a Monarch Butterfly"

The monarch butterfly is a handsome and heroic migrator. It is a flamboyant transformer: an egg hatches into a white, yellow, and black-striped caterpillar; then a metamorphosis takes place inside its leafy-green chrysalis, which is endowed with gold spots; the adult butterfly that emerges flaunts orange and black (fig. 1.1). In the monarchs' annual migratory cycle—perhaps the most widely appreciated fact about them—individual butterflies travel up to five thousand kilometers (three thousand miles), from the United States and Canada to overwintering grounds in the highlands of Mexico. After four months of rest, the same butterflies migrate back to the United States in the spring. Come summer, their children, grandchildren, and great-grandchildren will populate the northern regions of America.

But there is much more to the monarch's story than bright coloration and a penchant for epic journeys. For millions of years, monarchs have engaged in an

FIGURE 1.1. The monarch butterfly in three stages: (a) a caterpillar eating a milkweed leaf, (b) a chrysalis undergoing metamorphosis, and (c) an emerging butterfly before it expands its wings.

evolutionary battle. The monarch's foe in this struggle is the milkweed plant, which takes its name from the sticky white emissions that exude from its leaves when they are damaged. The monarch-milkweed confrontation takes place on these leaves, which monarch caterpillars consume voraciously, as the plant is their exclusive food source. Milkweeds, in turn, have evolved increasingly elaborate and diversified defenses in response to herbivory. The plants produce toxic chemicals, bristly leaves, and gummy latex to defend themselves against being eaten. In what may be considered a coevolutionary arms race, biological enemies such as monarchs and milkweeds have escalated their tactics over the eons. The monarch exploits, and the milkweed defends. Such reciprocal evolution has been likened to the arms races of political entities that stockpile more and increasingly lethal weapons.

This book tells the story of monarchs and milkweeds. Our journey parallels that of the monarch's biological life cycle, which starts each spring with a flight from Mexico to the United States. As we follow monarchs from eggs to caterpillars, we will see how and why they evolved a dependency on milkweed and what milkweed has done to fight back (fig. 1.2). We will discover the potency of a toxic plant and how a butterfly evolved to overcome and embrace this toxicity. As monarchs transition to adulthood at the end of the summer, their dependency on milkweed ceases, and they begin their southward journey. We will follow their migration, which eventually leads them to a remote overwintering site, hidden in the high mountains of central Mexico. Along the way, we will detour into the heart-stopping chemistry of milkweeds, the community of other insects that feed on milkweed, and the conservation efforts to protect monarchs and the environments they traverse.

To be sure, this story is about much more than monarchs and milkweeds; these creatures serve as royal representatives of all interacting species, revealing some of the most important issues in biology. As we will see, they have helped to advance our knowledge of seemingly far-flung topics, from navigation by the sun to cancer therapies. We will also meet the scientists, including myself, who study the mysteries of long-distance migration, toxic chemicals, the inner workings of animal guts, and, of course, coevolutionary arms races. We will witness the thrill of collaboration and competition among scientists seeking to understand these beautiful organisms and to conserve the species and the ecosystems they inhabit.

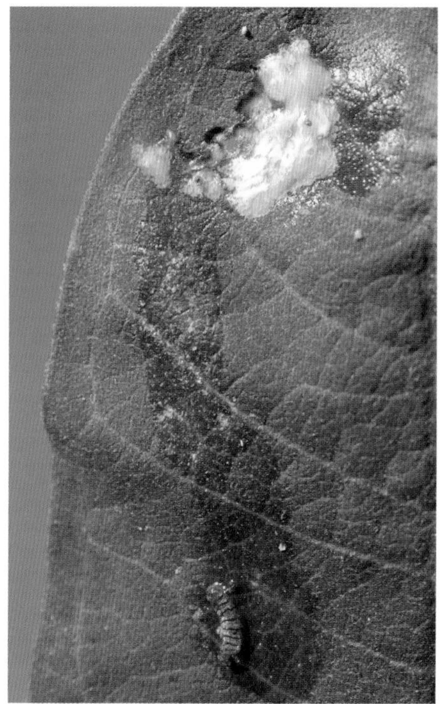

FIGURE 1.2. An unlucky monarch butterfly caterpillar that died after taking its first few bites of milkweed, the only plant it is capable of eating. In a violent and effective defense, toxic and sticky latex was exuded and drowned the caterpillar. A substantial fraction of all young monarchs die this way.

FROM SIMPLE BEGINNINGS

From a single common ancestor, milkweeds diversified in North America to more than one hundred species. And the monarch lineage is no slouch, with hundreds of relatives we call "milkweed butterflies" throughout the world. Although monarchs are perhaps best known in the northeastern and midwestern United States, they occur throughout North America, and self-sustaining populations have been introduced to Hawaii, Spain, Australia, New Zealand, and elsewhere (fig. 1.3). Interactions between butterflies and milkweeds now occur throughout the world, but this account focuses primarily on what happens in North America. The reason is quite simple: eastern North America is where the monarch (*Danaus plexippus*), considered by most to be the pinnacle of milkweed butterflies, coevolved with milkweeds.

The monarch's annual cycle in eastern North America involves at least four butterfly generations, with individuals crossing international borders several times. In spring, butterflies migrate from Mexico to the southern United States. Flight is fueled by nectaring on flowers and is punctuated by laying eggs on milkweeds. To grow and sustain each generation, milkweed is the only food needed. Three cycles—from egg to caterpillar, to chrysalis, to butterfly—occur as monarchs populate the northern United States and southern Canada each summer. And while nearly all mating, egg-laying, and milkweed eating occurs in the United States and Canada, each autumn monarchs travel to Mexico. At the end of summer, southward migrating monarchs fly thousands of kilometers and then rest for some four months before returning to the Gulf Coast states in the following spring. How and why they do it is a story that continues to unravel, and it no doubt will keep scientists busy for centuries.

The energy that builds a monarch butterfly's body ultimately comes from plants—as it does for all animals. For most butterflies and moths (collectively, the Lepidoptera), the caterpillar stage is essentially a leaf-eating machine. Perhaps it is not surprising then, that caterpillar feeding has led to the evolution of armament (or "defenses") in plants. The leaves of nearly all plant species are not only unappetizing to most would-be consumers; they are downright toxic. Milkweed's toxicity has long been known, and foraging on milkweed has surely killed countless sheep and horses. Most other animals avoid this milky, sticky, bitter weed, and yet monarchs came to specialize on it. While the toxic principles of milkweed keep most consumers at bay, monarchs and a few other insects have craftily adapted to the plant. Humans have used the chemical tonic of milkweeds as medicine for centuries, and so too have monarchs exploited their medicinal properties—at least at low doses. As the great Renaissance scientist Paracelsus noted five hundred years ago, "dose makes the poison"; there is often a fine line between poison and medicine. Much of this book is devoted to unraveling the evolution of poisonous and medicinal properties in plants that are habitual fodder for animals.

The evolutionary war waged between monarchs and milkweed is a product

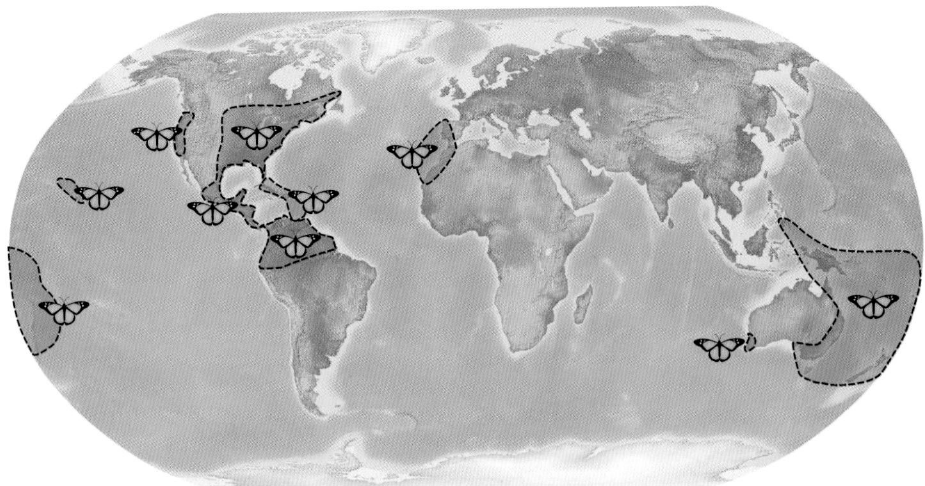

FIGURE 1.3. The worldwide distribution of monarch butterflies. Although native to the Americas, they have been introduced to the South Pacific, Australia, and Spain over the past few hundred years. The introduction of weedy milkweeds to these new regions, mostly the tropical milkweed *Asclepias curassavica*, preceded the establishment of monarchs. Monarchs are most abundant in North America.

of their intimate relationship. Monarchs not only tolerate milkweed's toxicity but have evolved to put it to work. For more than a century, insect enthusiasts have observed that most bird predators leave monarchs alone, presumably because their bright coloration signals a toxic body. Nonetheless, monarchs are not free of enemies. Flies and wasps consume them from the inside and eventually burst out. Tiny protozoan parasites infect their bodies, and monarchs medicate themselves with milkweed's toxins.

The milkweed plant is not a passive victim being devoured by monarchs. When the plant is attacked, its entire physiology, expression of genes, and toxicological apparatus kicks into high gear like an immune system. Milkweeds may lack an animal's central nervous system, but they possess all the other attributes common to the sessile sugar factories we call plants. They actively engage in strategies that defend against, tolerate, and when possible, manipulate insect enemies like the monarch butterfly.

While some mysteries of monarchs and milkweeds were only recently solved, much of what I present about the interaction between monarchs and milkweeds was reasonably well known (or at least hypothesized) more than a century ago (fig. 1.4). Searching through old newspapers, one can find beautiful accounts of their relationship. Although the classification of the monarch butterfly has changed over the past 150 years, the intimate interaction with milkweed was observed from the very beginning. Monarch "plagues" have been reported for at least as long, frightening entomophobes (people afraid of bugs). Nonetheless, because milkweed is sometimes considered an undesirable weed, an abundance of monarchs was also said to be beneficial by entomologists who knew the insect, as it might control the plant. There were newspaper reports of "Monarch Invasions from Canada" (as they migrated south past Rochester, New York) as early as the 1880s. Although there was some controversy about whether the butterflies migrated long distances, it was solidly hypothesized early in the twentieth century that this insect followed the seasons, south in the autumn, and with multiple generations moving northward each spring. How and why they migrate, and how and why they feed exclusively on milkweed, were discoveries made over the next hundred years. Honestly, they are not fully solved mysteries, but we have made great progress, and this book is about revealing the science behind these discoveries.

Monarchs have also been proposed as a sentinel, whose health as a species may be a "canary in a coal mine" for the sustainability of the North American continent. They travel through vast expanses, tasting their way as they go. Although they tolerate milkweed poisons, they are highly susceptible to others, especially pesticides. Summer and winter climates are likely the key drivers of the monarch's annual migration: feed on spring and summer milkweed foliage, follow the season north as it is progressively unveiled, rest in the chill mountain air in winter. Their time in Mexico is delicately balanced between being physiologically active, but cool, not burning precious energy before spring arrives. Our changing climate is certainly affecting monarch butterflies, although we are just beginning to understand the severity of these effects.

MYSTERY OF MIGRATION.

[St. Paul Dispatch.].

NCE again on their annual pilgrimage, which never has failed in the last 10 years, thousands of beautiful monarch butterflies, all the same color and all the same species, swarmed on the limbs of trees in the yard in the rear of the home of Mrs. W. D. George, Dayton avenue, yesterday afternoon, hanging in great clusters so thick in places the trees seemed spotted with brown.

About 3 o'clock they began to come, hundreds in a flock, flying as straight as if guided by an unseen hand. Past hosts of other trees they flew unhesitatingly, and with an unerring instinct sought resting places on the limbs of the very tree that has been the temporary shelter of thousands of their ancestors in years gone by.

More singular was the quest of the butterflies of former years that had sought out this resting place over night on their journey south. Nine years in succession—so long has Mrs. George kept account—they came to one particularly long limb of a box elder in the yard, literally concealing it with their clusters.

Other limbs on that tree were ignored as the butterflies yesterday ignored other trees in adjoining lots. This was their resting place for the night—no other would do. Unerringly they found it, hanging low to the ground, and never did they leave it until sunrise.

Last year this limb was cut off, so when the butterflies returned this year the old place of vantage was gone and they clustered on the outer branches of this and other trees in the yard. But in the one yard they stayed, loyal though the old home was gone.

What guides them to one spot in St. Paul is a scientific mystery made more involved because it is said butterflies live only one year, hence none from last year could have guided them this season to the trees. Mrs. George, who has watched them for years, and has seen thousands come and thousands go, offers no explanation for the ever recurring visit.

"This butterfly is known commonly as the milkweed butterfly," said A. G. Ruggles, assistant professor in the division of entomology of the State Agricultural School, when consulted last night. "The scientific name is Anosia plexippus, and it belongs to the family Lymnadidas. It derives its common name because the caterpillar from which it comes feeds almost entirely on milkweed.

"These butterflies are nauseous to birds and hence are able to congregate as they do with impunity. They have a northern migration in the spring and a southern in the fall, different from that of birds in this way, that the female butterflies come north about the time the milkweed begins to grow in the spring, lay their eggs and die soon after. Hatching from these eggs are the caterpillars that feed on the weed, eventually changing into butterflies, which fly still further north, where the females again lay eggs.

"And so generation after generation is produced, the adults migrating further and further north, until we find them late in the season often as far north as Hudson Bay. In the fall they gather in great flocks and fly south.

"They are great fliers, and we know they can cover 500 miles at least, because specimens have been taken in the ocean that distance from shore. The monarch is a native of America, but is now practically cosmopolitan, being found almost everywhere on the globe."

FIGURE 1.4. A newspaper article about the monarchs' migration from the *Washington Post*, September 17, 1911.

In some respects, human activities have enhanced habitat for milkweeds and monarchs north of the overwintering grounds. Logging and agriculture have been good for monarch populations in some regions, like the eastern United States, where these pursuits likely made milkweed and its associated butterflies much more abundant. However, farming surely destroyed much of the midwestern prairie, where milkweed had previously been prolific. Now the same processes, combined with the indirect influences of other human activities, have been suggested as drivers in the decline of monarch butterfly populations. I evaluate what is known about the causes of monarch and milkweed ups and downs toward the end of this book. If they are truly sentinels, then much more than the sustainability of monarchs is at stake, and careful study of their biology—past, present, and future—is in order.

GETTING INFECTED

How ecological interactions—plants and insects, monarchs and milkweeds—caught my attention is a story in and of itself. I grew up in a fairly rural area of suburban Pennsylvania, where fields of red clover and foxtail grasses were common, and my brother and I were encouraged to spend much of our time outside. Vacations were spent camping; my mother was, and continues to be, an insatiable gardener; and the corn fields growing behind my home prompted me to want to be a farmer. As a college student at the University of Pennsylvania, I felt the bliss of self-discovery, yet also the pressures of being a child of immigrant parents who were unfamiliar with most academic endeavors outside of medicine and engineering. My parents' proviso concerning my college education was that, in addition to exploring my interests in social science and the humanities, I take introductory science and math classes, so as not to close too many doors. Fair enough.

As a sophomore, I decided to take introductory biology. But, because the lecture halls were limited in their seating, and because many colleges feel pressure to have smaller classes (after all, small classes enhance students' learning, as well as college rankings), there were two offerings of the course that semester—similar classes, covering much the same material, but taught by different professors. To choose, I did what many students did, and still do: I consulted what was known as a "skew guide," a "for the students, by the students," survey of courses that outlined the degree of difficulty, what was liked and disliked by students who had taken the course previously, and unashamed caricatures of the esteemed faculty—the clothes they wore, comments about their hygiene, and notes about their traits, usually having little to do with their ability to impart scholarly information. Sad but true, what sealed the deal for me was the characterization of one of the professors: "typically comes late to class and leaves early." I actually don't remember if that ended up being true, but the course, and his approach to biology, caused a profound shift in my own development as a student. Dr. Daniel Janzen presented biology as a set of stories, far

FIGURE 1.5. In a memorable lecture on butterflies and poisonous plants, Dr. Janzen showed this slide depicting the unpalatability of milkweed. The toxic plant flourished under grazing pressure because it was avoided by horses (*right side of the fence*). But where horses were absent (*left*), milkweed was less abundant and suffered from competition with grasses.

stranger than any science fiction I had read. Biology was a series of mysteries that could be solved by careful observation and clever manipulation. Biodiversity was presented as a bottomless mine of species and interactions that had been shaped by both millions of years of evolution and the now dominant species on the planet, *Homo sapiens* (fig. 1.5).

A recurring theme of the course and the professor's favorite organisms for study were plants that were damaged by insects. Insects eating plants? What about the charismatic megafauna: lions, tigers, and sharks? Or at least buffalos and birds? Now that I am a professor, I annually teach a course called "Chemical Ecology" with several faculty colleagues at Cornell University. In this course we analyze how chemicals in the natural world mediate interactions between species. Why are chilies and horseradish spicy? How do monarch butterflies gain their toxicity? And, are there really human pheromones? Yet, our lectures often focus on insects eating plants. Comments from students in our course evaluations occasionally plead, "Enough with the caterpillars already!" Yet it is the abundance, diversity, and general importance of insect-plant interactions that motivate our course, as well as my own fascination and research focus on monarchs and milkweeds.

ARTHROPOD-PLANT INTERACTIONS

What can little creatures like monarch butterflies and their vegetarian habits teach us about nature? First, the source of essentially all of the energy that powers an animal—really, any food chain—comes from plants. They constitute an exclusive group of organisms that can process nonliving matter and turn it into the energy that is needed for life. That process is photosynthesis, and that energy is sugar. Plants take sunlight, carbon dioxide from the air, and water, and through a chemical reaction produce oxygen and sugar. That sugar powers life on earth. Sure, we don't typically think of lions, tigers, and sharks as relying on plants. But they do. They eat other animals that survive by eating plants. Meanwhile, plants "eat" earth, wind, and sun. As such, plants make up the largest fraction of living matter (what biologists call "biomass") on the planet. Milkweed is but one of hundreds of thousands of plants species, yet it is an excellent representative to teach us about biology.

$$6CO_2 + 6H_2O + \text{photons} \longrightarrow C_6H_{12}O_6 + 6O_2$$
$$\text{carbon dioxide} + \text{water} + \text{light energy} \longrightarrow \text{sugar} + \text{oxygen}$$

Second, there are two pathways by which plant energy enters a food chain: as compost and as salad. Most of it, probably 80 percent or so, enters the food chain as rotting compost. As leaves fall off plants, microbes, worms, and microarthropods shred it, transform it, and make it available as broken-down food and nutrients for others. The rest, about 20 percent of plant material,

FIGURE 1.6. *The Bioscape*, where taxonomic groups are drawn proportional to the number of currently known species. In this diagram, the monarch represents all insects. Among the 2 million described macroscopic species (visible to the naked eye), about one quarter are herbivorous insects. Yet, our best guesstimates of the actual number of herbivorous insects on the planet range from 2 million to 5 million species. As we discover the rest of the species, plants are likely to constitute about 10 percent of the total number of species. Even by the most generous estimates, all vertebrates combined (including mammals, birds, fish, reptiles, and amphibians) would hover around 2 percent of species, and 70 percent of species are likely insects (about half of which are herbivorous). Accordingly, in terms of the source of the planet's overall energy, biomass, and biodiversity, plants and herbivorous insects play a dominant role.

enters the food chain fresh, as a "salad." Monarchs are but one of the millions of leaf-eating species that can teach us about the consumption of living plant tissues by animals. And, yes, although it is true that zebras and porcupines are charismatic megafaunal herbivores (big mammals that eat plants), in other ways, insects like monarchs dominate the scene (fig. 1.6).

The third reason for a focus on insects and plants is quite practical. These organisms are typically easy to cultivate in large numbers, in relatively small spaces, and with relatively little interference from animal rights activists. The consequence for scientists is that we can work toward strong inference in our studies. "Strong inference," a term introduced by a biophysicist and philosopher of science, J. R. Platt, in 1964, refers to going beyond single factors as explanations of natural phenomena, and going beyond correlations as explanations for the causes of patterns. Many correlations are statistical associations that have no "causal" basis. The variable on the *x* axis of a graph, although correlated with a variable on the *y* axis, is not the cause of variation in the variable on the *y* axis. Take, for example, the strong positive correlation between the per capita consumption of chocolate in a country and the number of Nobel laureates from that country. To quote from a tongue-in-cheek article published in the *New England Journal of Medicine*: "[We] estimate that it would take about 0.4 kg of chocolate per capita per year to increase the number of Nobel laureates in a given country by 1." To evaluate correlations rigorously, and hence to promote strong inference, one typically needs large numbers of study subjects, the testing of alternative explanatory factors, and a critical experiment. The abundance of insects on plants, along with the ability to raise them in both the laboratory and field, makes them ideal scientific subjects. Distinguishing between correlation and causation is critical to our understanding of the biology and conservation of monarchs and milkweeds. Turing back to our study of chocolate: countrywide spending on science also correlates with per capita income, the latter of which correlates with chocolate consumption (at least in the Western world). Even so, I would happily participate in a controlled study to determine the influence of chocolate consumption on scientific discoveries.

Given the attributes of insects and plants, it is perhaps not surprising that their study has become a bit of cottage industry among biologists. We even have a specific journal dedicated to publishing scientific studies on insect-plant interactions. Well, to be honest, it is called *Arthropod-Plant Interactions*, so as to include studies of related creatures with more than six legs, such as spiders, mites, millipedes, and centipedes. The main point, however, is that in addition to their general abundance, diversity, fascinating biology, and tractability of study, these insect-plant systems have become a general model for understanding biology. Given the millions of species on the planet, biologists make progress through the in-depth study of several "model systems" that intensely examine a few selected organisms, with the hope of generalizing to other systems. Sometimes these "models" are specific species (like the lab rat). In other cases, however, the models may be habitats or groups of species that have their own ecologies.

Monarchs and milkweeds have proven to be an excellent model system through which we can understand the coevolution and conservation of species. Like all plants, milkweeds have an arsenal of toxins, evolved by natural selection, to ward off pesky herbivores. Like all herbivores, monarchs have a diverse portfolio of tolerances and strategies that leave them undeterred from feeding on their food. And all animals have their own enemies, predators, parasites, and microbial infections. Monarchs are no exception, and yet the detailed study of their relationships has revealed a role for milkweed's toxic properties in the interaction between monarchs and their enemies. The monarchs' spectacular form and flight, although extravagant, demonstrates the lengths to which natural selection can go. And because of the food and habitat needs of the monarch along its annual migratory cycle, monarchs have now become important in understanding general principles of species conservation. Monarchs and milkweeds have served as an important icon for debates and concerns about genetically modified organisms, climate change, and environmental issues more broadly.

FINDING MONARCHS AND MILKWEEDS

Inspired by Professor Janzen's introductory biology class, fueled by a newly found passion for studying insects on plants, and advised to find one of the big state universities ("the land grant colleges," as they are called), I ended up pursing a PhD at the University of California, on the Davis campus. I landed at UC Davis probably because of the confluence of several factors: it was far enough from home that it sparked some sense of exotic mystery; it provided the opportunity to study with a talented and beloved mentor, Dr. Richard Karban, who remains a friend and inspiration; and it had one of the top programs in ecology and evolutionary biology. My parents were slightly befuddled, if not worried, by this choice. The university was not a big name school, it was far from home, and, after all, I was leaving the traditional lines of inquiry to become some sort of bug doctor.

It was not until near the end of my time at Davis, however, when my wife, Jennifer Thaler, and I had both secured faculty jobs in the Botany Department at the University of Toronto, that a path first led me to the world of monarchs and milkweeds. As I looked forward to starting new research projects at a new job, I found it both an exciting and daunting challenge. How to pick a fruitful set of organisms to work on, one that would provide scientific insights, prove amenable to discovery, and perhaps above all, promote inspiration? My friend, collaborator, and graduate student colleague, James Fordyce, known to most as Jimmy, and to some as Uncle Jimmy, suggested that I consider working on monarchs and milkweeds (fig. 1.7). Now a professor at the University of Tennessee, he had worked on milkweed insects for his master's degree. The reasoning behind Jimmy's proposal is worth explaining, because it directly relates to what makes monarchs and milkweeds ideal study subjects.

The first attribute of monarchs and milkweed is that they have long been the recipients of love and affection from all kinds of people. And this love is expressed is various ways: by observations in nature, hands-on husbandry, the buying and selling of butterflies and seeds, digital information, symbols in the logos of organizations, and scientific study. Hundreds of thousands of monarchs are reared each year in classrooms and homes in the United States alone. At least as many are reared by butterfly breeders for sale and releases at weddings and other special occasions. Any search of "monarch butterfly" on the Internet yields hundreds of thousands of websites. Over the past twenty years, the *New York Times* alone has published more than one hundred articles about monarchs, averaging one every two to three months. Monarchs have become logos and symbols for organizations far and wide, including the Union of Con-

FIGURE 1.7. A typical summer sight in eastern North America—a monarch caterpillar getting ready to feast on the common milkweed, *Asclepias syriaca*.

cerned Scientists, many K–12 schools, the non-GMO project, and numerous corporations whose businesses range from manufacturing to banking. And, finally, a recent survey of thousands of Americans revealed that, collectively as a nation, households are willing to donate nearly $5 billion to aid in the conservation of monarch butterflies. The study revealed a willingness to donate on par with many endangered vertebrate species (although not as much as for bald eagles, elephants, and gray whales).

As a result of public interest and much scientific study, we know a tremendous amount, not only about these two organisms' natural history, but also about their coevolutionary basics. Common milkweed was transplanted and established in Europe as a possible (failed) source of rubber for tires and fluffy fill for pillows. Milkweeds are toxic plants that most animals do not eat. Monarchs cannot live without the milkweeds; it is the only food they eat. But milkweeds and monarchs are both toxic. The plant poisons, which have been used medicinally for hundreds of years, are taken up by monarchs and ward off bird predators. The monarch migration is legendary, mind-boggling, and indefatigable. The list goes on. This plant-insect interaction brings together a renowned butterfly, attractive plants, and a rich vein of history, biology, and ecology.

Second, what makes monarchs and milkweeds good scientific subjects is advertised by the plant's common name. How I have lamented the "weed" in the name "milkweed." As Ralph Waldo Emerson once quipped, a weed is simply "a plant whose virtues have not yet been discovered." Other previously used names for this beautiful plant include swan plant and silky swallow-wort, but none other than milkweed stuck. Nonetheless, the weediness, at least of common milkweed (*Asclepias syriaca*), and the weediness of monarchs themselves, is an attribute that makes them great biological subjects because they are abundant. The insects and plants alike are easy to spot. The butterflies are colorful, typically not elusive, and active during the daylight hours. These species make the process of doing science—rearing lots of them, screening them for their traits and behaviors, finding them in the field, and grinding them up in the laboratory, an easier process. Although working on rare species is certainly an important task for ecologists, rarity presents a set of challenges before scientific investigation even begins.

Third, both monarchs and milkweeds are native to North America. Biologists are often obsessed with this notion of "native," because it is thought to reflect some primordial state untouched by humans. Of course, nothing could be further from the truth. Even in the depths of the Amazon or Siberia, humans have had an impact on most organisms and the habitats they occupy. Nonetheless, the native state of both partners in an ecological and evolutionary relationship makes it such that their behaviors and their biology were potentially shaped by their long-term interactions with one another. And monarchs and milkweeds do share a deep evolutionary history; they have existed together for a very long time, likely millions of years. Thus, we can interpret the ecology of this system through the lens of natural selection and coevolution.

And finally, monarchs and milkweeds are a natural system with some balance between complexity and simplicity. Complexity can mean many things, but here I am thinking about diversity—of species, habitats, and interactions between species. Monarchs come from a group of some 6,000 brush-footed butterflies (in the family Nymphalidae), and many of the smaller grouping of 170 "milkweed butterflies" (in the tribe Danaini) interact with milkweeds (fig. 1.8). As I will discuss later, monarchs have an intimate association not only with milkweeds but also with microbial parasites, some of the other insects that eat milkweed, and a whole community of predators, from birds to spiders, and from stinkbugs to wasps. Not only does this diversity of potential interactions set the stage for endless scientific study, but the monarchs' yearly travels expose them to substantial variation in when and with what they interact. It is this diversity of species, interactions, and environments that they live in that is food for scientists: mysteries to be solved. Yet, for groups that are much more diverse, or associated with wildly different plants and parasites, the complexity can be overwhelming and make scientific progress quite slow.

The complexity-simplicity balance also applies to the plants. Milkweeds come from a genus with about 130 species (given the genus name *Asclepias* by Carolus Linnaeus after the Greek god of medicine and son of Apollo, Asklepios). All *Asclepias* live in the Americas, with most living in Mexico and northward. It is not a tropical group of plants. The evolutionary sister group to *Asclepias* are more than 250 species in an African genus called *Gomphocarpus* (fig. 1.9). Of the American milkweeds, most are rare, and only a few species, like common milkweed, *Asclepias syriaca*, are highly abundant and noticeable in many environments. Species of milkweed do, however, inhabit some of the most diverse habitats available, from standing water to the driest of the dry deserts. And although most *Asclepias* live in open habitats, preferring full sun,

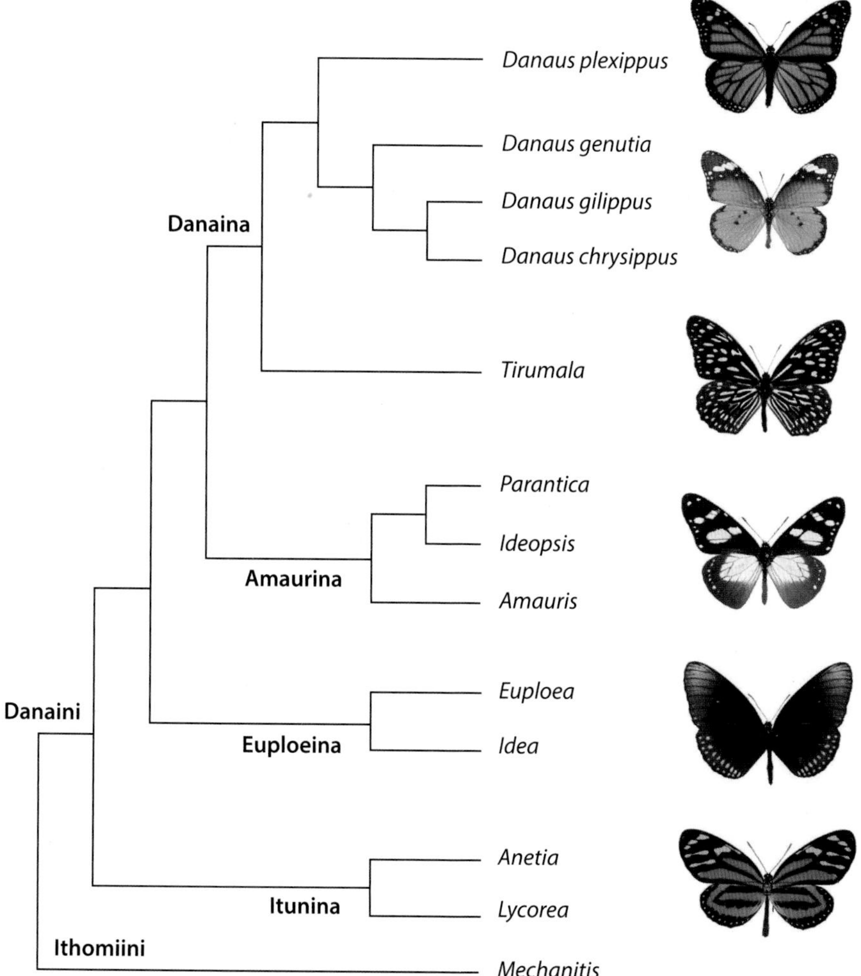

FIGURE 1.8. The monarch arose from a group of about 170 physically similar, yet evolutionarily distinct species known as the milkweed butterflies (tribe Danaini). Shown is a phylogeny, or visual representation of the evolutionary relationships in this tribe. Note that most milkweed butterflies are not shown here (for example, there are twelve recognized species in the genus *Danaus*). Instead, representatives of each of the major groups are shown. The Ithomiini is the sister tribe to the Danaini and contains several hundred tropical species.

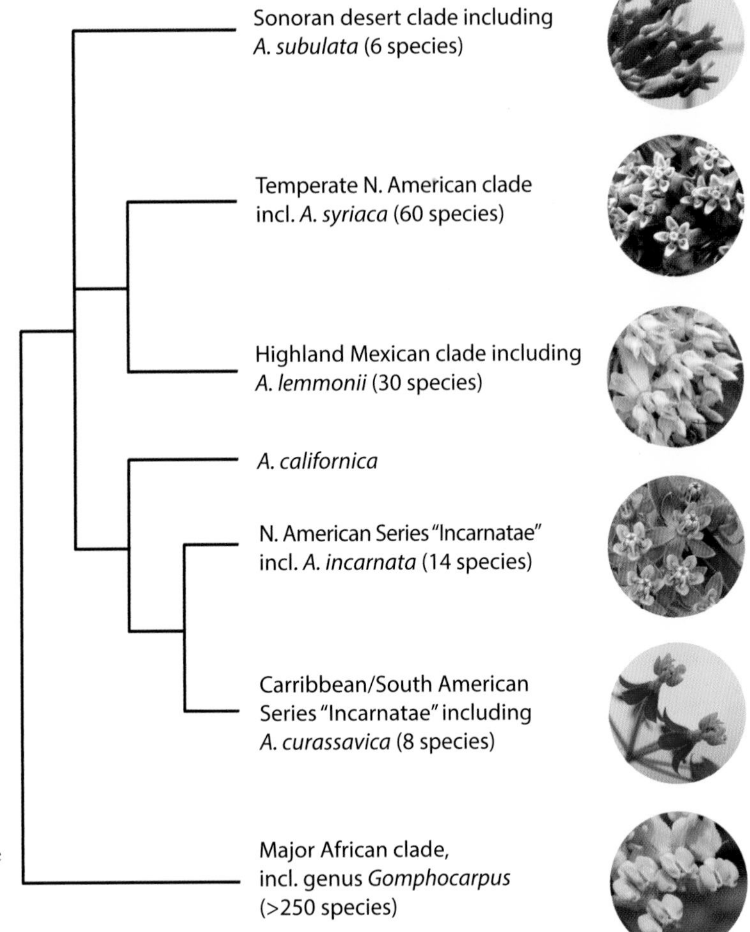

FIGURE 1.9. A summary phylogeny showing the evolutionary relationships of the milkweeds in plant genus *Asclepias* (with about 130 recognized species in North America), here trimmed to show representatives of the major groups. No image of *Asclepias californica* is shown, and the representative of *Gomphocarpus* is *G. fruticosus*.

a few species inhabit the forest shade. All are herbaceous (not woody) and perennial. Together these attributes point to a level of manageable complexity. That is, the plants have evolved from a single ancestor into many species, into many habitats, and with some variation in their ecology.

Like all plants, these evolutionarily related species of milkweeds have their own special community of herbivores, giving the plants a predictable and well-defined group of insect enemies. And this too is an attractive attribute of the milkweeds for scientific study. Take the common milkweed, which is fed upon by eleven insects: three aphid species suck the phloem sap, two lygaeid bugs eat seeds, three different beetle species bore through the roots, tunnel in the stems, or eat the leaves, and a small flattish fly mines between leaf layers (see figures in chapter 7)—not to mention a moth and a butterfly species whose caterpillars chew the leaves. The insect community is complex yet simple. The complex part is that these insect herbivores have divvied up the plant, with different species eating different plant parts. Also, these insects span a tremendous taxonomic breadth, covering some 350 million years of evolutionary history. What I mean by this is that the diverse species of insects that now eat milkweed shared a common ancestor about 350 million years ago—the proto-insect. Over the past hundreds of millions of years, the insects' ancestors rampantly diversified, giving rise to honeybees, mosquitos, beetles, and butterflies, and many times independently distinct groups of these insects would colonize and adapt to eating milkweeds. Interestingly, these insect species are essentially all specialists. They are not omnivorous. They are not even adventurous. All they eat is milkweed. And therein lies the simplicity of monarchs and milkweed. The insects are confined to feeding on milkweed, and therefore we know where to find them; their dietary habits are well-defined; and they are decidedly pests on the plant.

Our journey will now start in earnest by placing monarchs and milkweeds in the context of their stockpiling arms race. The nature of the monarch-milkweed interaction is simplified by the fact that monarchs are unquestionably pests and are not also pollinators or beneficial in any other way. Milkweeds must defend themselves. Early studies of this interaction led to the birth of a new scientific discipline called chemical ecology, which among other topics, tries to decipher the mechanisms and consequences of such arms races. Now that the field is maturing, it has created new scientific questions. What is most fascinating about monarchs is that they have cracked the milkweed's code of defense, forever changing the course of their coevolutionary interaction.

CHAPTER 2

The Arms Race

> There is grandeur in this view of life, with its several powers, having been originally breathed into a few forms or into one; and that, whilst this planet has gone cycling on according to the fixed law of gravity, from so simple a beginning endless forms most beautiful and most wonderful have been, and are being, evolved.
>
> —Charles Darwin, *On the Origin of Species*, final sentence

A monarch butterfly is perched on a milkweed flower, ready to take a sip of nectar. What a site of harmony in nature (fig. 2.1). But this butterfly is not a pollinator of milkweed. Instead, the butterfly is hoping to find a mate and then to have children that will devour the milkweed plant. Through evolution, the butterfly has adapted to exploit the plant, but there is nothing in it for the milkweed. And through a process called coevolution, the milkweed does not invite the monarch, but rather tries to ward it off. Their battle has been intense, so much so that back in Darwin's day, scientists used monarchs and milkweeds to advance new ways of looking at nature. Out of these insights, new disciplines of science emerged, including what we now call chemical ecology. What scientists eventually discovered was that the monarch-milkweed relationship initiates an extraordinary arms race.

EVOLUTION AND COEVOLUTION

All organisms on the planet, from ants and bacteria to cats and dogs, to elephants, figs, and giant saguaros, descended from a single, universal common

FIGURE 2.1. A male monarch butterfly perched on the flowers of common milkweed in the author's front yard.

ancestor. Darwin got it right when, in the last paragraph of his 1859 opus, *On the Origin of Species*, he wrote about this common ancestor and the subsequent and sustained evolution of diverse forms. This represents what is termed "macroevolution," that which involves the generation (and loss) of distinct species. Such evolution is evident all around us, and many biologists spend their time interpreting the features of life in the context of evolution, because that is the only context in which biology makes sense.

The mechanism of Darwinian evolution is what we term "microevolution." Microevolution is a change in the frequency of alternate forms of genes (or alleles) within a population of a single species, where these genes code for traits presumably important to the organism's form and function. Through natural selection, the frequency of particular genes in a population changes. Think about the four human blood types (A, B, AB, and O), which are determined by a single gene. In some human populations, natural selection has favored a higher representation of particular forms of that gene, typically because different blood types are more or less resistant to particular diseases.

In their simplest forms, micro- and macroevolution are beautiful, powerful, and commonplace at the same time. Polar bears' white fur helps to camouflage them in the arctic, while their underlying black skin absorbs heat. Plants living in deserts have evolved many ways to conserve water, and animals that live in caves commonly lose their eyesight over evolutionary time. But evolution occurs not only with respect to the physical and chemical attributes of the environment (such as moisture, light, and temperature), but also in relation to the "biotic environment," the other organisms with which a species interacts. Take, for example, the evolutionary origin of the revolutionary drug penicillin.

Fungi that eat dead and decaying things, like the *Penicillium* molds that gave us penicillin, evolved to produce antibiotics to defend the resource they are eating from other tiny consumers, like bacteria. Each of the hundreds of species in the genus *Penicillium* produces distinct chemicals, most of which have been shaped by natural selection. Natural selection can be likened to a filter. As long as there is heritable variation in a population, natural selection can cause evolutionary change, because those individuals with advantageous traits are the ones who survive and successfully reproduce. Natural selection hones traits within populations of a species, and as populations diverge, macroevolution takes hold, allowing for speciation. In the case of *Penicillium*, diverse antibacterial toxins have evolved as a defense among the many species in this genus of fungi.

Coevolution takes evolution in response to the biotic environment to the next level (fig. 2.2). When organisms interact, like the competing molds and

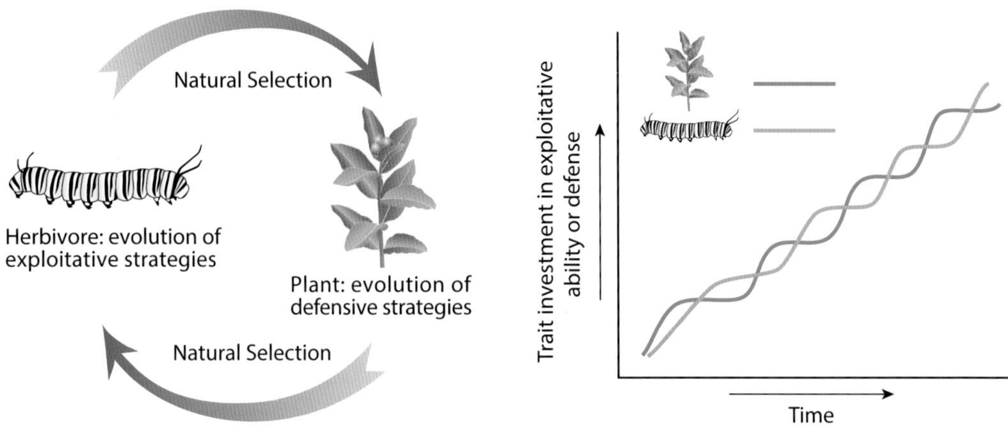

FIGURE 2.2. Two ways to envision coevolution. *Left*, reciprocal natural selection is illustrated as a continuous cycle. *Right*, the same process over time results in an "arms race," with increased investment by the plant in defensive traits and by the insect in traits that exploit the plant. Note that there are other possible outcomes of coevolution not discussed here.

bacteria, or the mutually beneficial relationship between flowering plants and pollinating bees, or the cat-and-mouse game of predators and prey, coevolution is possible. Coevolution is the reciprocal adaptation that occurs as species interact. The term "adaptation" means that the frequencies of particular beneficial genes in a population will increase. In coevolution, the changes are imposed by the interaction partner, say the cat imposing natural selection on the mouse. The evolutionary response to this selection may be that mice are better camouflaged or faster runners, as the more apparent or slower individuals will be removed from the population by predators. Critically, what makes coevolution different from everyday evolution is that coevolution involves reciprocity. That is, following adaptation by the mouse, say with faster and faster running speeds, natural selection is now imposed on the cat, causing the evolution of a more acute ability to sniff out or chase down mice.

The birth of coevolution as a concept, studied as the warfare between plants and herbivores, is credited most prominently to Ernst Stahl, who summarized his studies of snails and plants in 1888, only a few years after Darwin passed away. Stahl developed a theory of coevolution, and his framework set the stage for unraveling some of the mysteries of monarchs and milkweeds. His critical experiment was to take plant species that appeared to be avoided by snails in the field, to show that in the laboratory snails refused to eat these plants, and then to remove chemicals from the leaves using an extraction procedure (imagine soaking the leaves in alcohol overnight) and showing that those leaves were now palatable to the same snails. He interpreted his results in light of "reciprocal adaptation," a very early reference to what we now call arms race coevolution. In particular, Stahl noticed that some snails were generalist feeders, and their ability to eat plants was enhanced by removal (extraction) of the plant's chemical defenses. Alternatively, other snail species limited their diet to one or a few plant species—call them "specialist herbivores"—and these snails were observed to prefer plants in their intact state. It stood to reason that specialists were engaged in an arms race with the plants, eventually overcoming the plants' defenses, and even using these "defenses" for their own purposes.

MONARCHS, I DON'T REALLY NEED YOU

How species involved in interactions evolve and coevolve depends acutely on the nature of the interaction. If two species are strictly antagonists, then an arms race may ensue. Sometimes, however, species play dual roles, positive under some circumstances and negative in others. In this case, say if the monarch were both a beneficial pollinator and an herbivorous pest of the milkweed, perhaps the plant would not mount defenses. The benefits of butterfly pollination to milkweed could outweigh the costs of caterpillar herbivory. But here is where I must dispel a widely held myth about monarch butterflies. Milkweeds do not need monarchs, because the butterflies are simply no good as pollinators. Monarchs are strictly pests.

Unlike many other coevolutionary relationships, that between monarchs and milkweeds is not symbiotic. Although many definitions of symbiosis exist, nearly all require the relationship to be close, and many definitions require the relationship to be mutually beneficial. From the monarch's perspective, the relationship is intimate and beneficial. However, from the milkweed's perspective, it is neither. The importance of this distinction will be increasingly apparent as we move forward, but for now let's focus on the fact that monarchs frequently can be found collecting nectar from milkweed flowers. They drink nectar, but they are typically ineffective at pollinating the flowers. Without understanding this issue, we cannot understand the nature of the monarch-milkweed arms race.

First, a bit of the birds and the bees—the reproductive biology of milkweeds. All milkweeds are perennial (they typically live and reproduce for many years). Despite their disappearance aboveground in the winter, the bulk of their biomass lives underground in the soil, stays dormant in the winter, and has plenty of stored reserves to power new shoots in the spring. Additionally, some, although certainly not all, milkweeds are clonal. That is, they send stems foraging underground and pop up new shoots when needed. This cloning of stems is in part responsible for common milkweed's weediness. In addition, milkweeds reproduce sexually. They produce flowers, attract pollinators, send pollen to other plants, and receive pollen to fertilize ovaries and initiate the production of seeds (fig. 2.3). Most milkweeds do not accept their own ("self") pollen; they need to receive pollen from an independent milkweed plant in order to successfully make a fruit full of seeds. Milkweeds cover a lot of bases with this way of life. They live many years with many opportunities for reproduction, which after all, is the main goal of all life. Where conditions are good and a particular plant is successful, milkweeds clone themselves locally. But, as a means to colonize new habitats, and to mix genes with mates, milkweeds engage in pollination and sex.

Most important, the reproductive cycle of milkweeds does not include a role for monarchs. Although frequent visitors and drinkers of milkweed's nectar, monarchs are ineffective extractors and deliverers of pollinia (or pollen sacs). Milkweed flowers do not offer up loose pollen grains the way 90 percent of plants do. In fact, two groups of flowering plants, orchids and the subfamily that includes milkweeds (Asclepiadoideae) have evolved pollen packages called

a

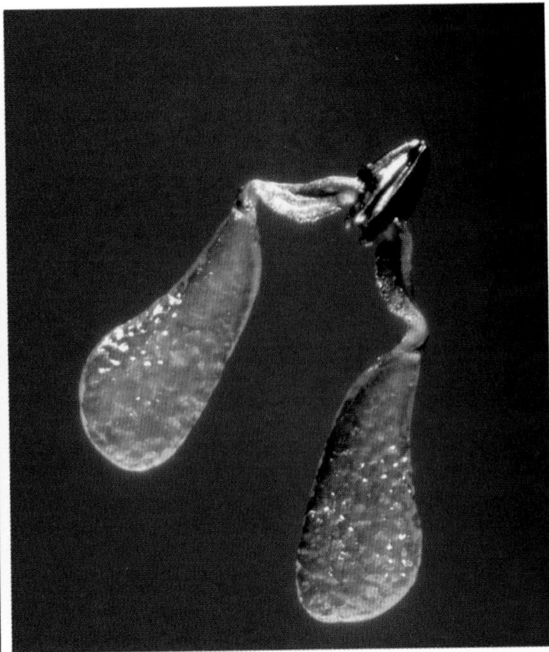

b

FIGURE 2.3. Pollination of milkweeds. When bees and other large insects in the order Hymenoptera visit milkweed flowers to drink nectar, (a) their legs slide near the flower's slit, and the bee's leg hairs often grab the top of the (b) wishbone-shaped pair of pollinia. (c) A successful removal of the pollinia results in its becoming attached to the bee's leg. Later, when the bee climbs over other flowers, a pollinium incidentally gets inserted into a slit, allowing

"pollinia," each with hundreds of tiny pollen grains. Milkweeds do not have loose pollen grains that can be collected by bees, rubbed onto insect abdomens, stuck to a butterfly proboscis (the tubular drinking mouthpart), or blown in the wind. So, for pollination to occur, the pollinia must be extracted from the milkweed flower and then inserted into the flower's female slit (or "stigmatic groove" in botanical terms). As my colleague Steven Broyles from the State University of New York at Cortland says, monarchs, with their long legs, simply don't contact the business-end of the milkweed flower.

In other plant-insect associations, the same insect species may serve as a pollinator and also as herbivore of the plant. Although this is most famously known from the yucca plant and yucca moths, and figs and fig wasps, many plant-

c

d

for fertilization of the ovules by germinating pollen grains. (d) Monarchs, however (shown here on *Asclepias tuberosa*), are not good pollinators of milkweed. Because of their large size and way of sitting on flowers, monarchs uncommonly come into contact with the pollinia and slit.

insect relationships follow suit. For example, the cabbage white butterfly (*Pieris rapae*), perhaps the most abundant butterfly in the world, not only pollinates mustards (relatives of cabbage in the family Brassicaceae), but also lays eggs on the plants after pollinating. Their larvae are voracious herbivores of mustard leaves and flowers. In such relationships, there is a conflict for the plant: how to attract and effectively use a pollinator without suffering from that pollinator's young consuming the plant. No such conflict exists in the milkweed-monarch matrix, so the plant can focus its energies first and foremost on defending against monarchs as pests.

This nonpollinating aspect of monarchs is not widely appreciated. Although monarchs may successfully pollinate some plant species (perhaps in the sunflower family, Asteraceae), this phenomenon has not been well-studied, and they are surely unimportant compared with the myriad other flower visitors. Nonetheless, in a recent presidential memorandum (June 20, 2014), "Creating a Federal Strategy to Promote the Health of Honey Bees and Other Pollinators," Barack Obama singled out monarchs as the only species other than the honeybee (*Apis melifera*) to be named as an important pollinator. Because the monarch is *not*, however, a good pollinator, the arms race proceeds, with plants evolving defenses to reduce attack by monarchs and insect adaptations to overcome these defenses.

EAT AND DON'T BE EATEN

The arms race signifies a battle between species, although not one where they are simply stockpiling weapons to destroy each other. In the arms race between plants and herbivores, plants evolutionarily accumulate defenses while the herbivore evolves means to circumvent these defenses. For milkweeds, the plant has evolved several forms of armament, including potent toxic chemicals, and monarchs have evolved the physiological means to tolerate these chemicals. As coevolution proceeds, escalation in defense and offense reciprocally ensues (see fig. 2.2).

But life isn't so simple. Every organism has its need for food and also has its own enemies to contend with. A major issue for most animals, really any organism, be it vegetable, animal, or germ, is to eat and avoid being eaten. These factors, along with finding mates and surviving abiotic stresses (such as extreme temperature, weather, and ultraviolet light) make up a large fraction of what we might call the ecology of an organism. Insects are no different: cope with the environment, eat, and don't be eaten. For insect herbivores, the food is decidedly vegetable, but what plant species to eat, what specific plant tissues to focus on, and having a sufficient supply, especially one that is not highly sought after by competitors, are important factors. Ditto for not being eaten. Insect herbivores may hide or avoid predators by crypsis (blending into the

Butterflies and Birds.

From Country Life in America.

Of all the "children of the air" that gladden a June day, the monarch butterfly is one of the most noticeable. Its wings shimmer like gold alloyed with copper as it pursues its lazy flight in the sunshine. The male monarch is a true dandy and carries on each hind wing a black sachet bag containing a strong perfume, most attractive to the other sex. The monarch is immune from bird enemies; the callow birdling that takes a bite from it wipes his beak in disgust and forever after connects the noisome taste with orange wings. A too hasty conclusion of which the Viceroy butterfly takes advantage and, by donning the monarch's uniform, escapes scatheless, although any bird might find it a beaksom morsel.

FIGURE 2.4. Newspaper article about mimicry from the *Washington Post*, June 8, 1902.

environment), or they may simply flee when they sense the risk of predation. However, some animals, like the monarch butterfly, eat in the open, do not blend in with their plant, and seem to advertise themselves through highly contrasting coloration. As we will see in later chapters, the coloration of monarchs is linked to their toxic diet. This is where the monarch-milkweed arms race took a turn, with monarchs taking advantage of milkweed's defensive poisons.

As early as the late nineteenth century, it was known that monarch butterflies were unpalatable, living with immunity from most predators, especially birds. A newspaper account from 1902 accurately describes not only their toxicity, but how bird predators learn to avoid the monarch's coloration and how the unrelated, and less toxic, viceroy butterfly benefits from coloration similar to that of monarchs (fig. 2.4). This phenomenon, termed Batesian mimicry, after the English naturalist Henry Walter Bates of the same era, will be discussed in chapter 6. What was less widely accepted at the time, but still studied among scientists today, is the origin of the butterfly's distastefulness.

YOU MAKE ME PUKE

In the late nineteenth century, a Belgian scientist, Léo Errera, was studying plant poisons called alkaloids, widely known for their pharmacological effects. Caffeine, capsaicin, morphine, and nicotine are all alkaloids, to name a few. In 1887, he made conclusions about their functional role in plants: "Most alkaloid-containing plants are avoided by browsing animals. A few grams of alkaloids are equally efficient protective means as the most forceful thorns." But what if in the course of a coevolutionary arms race, specialized herbivores not only overcame this toxicity, but evolved a means to put these chemicals to work for themselves? This hypothesis is termed "sequestration" and suggests that at a late stage in the arms race, some herbivores might be taking in and storing toxic compounds from their host plants. Sequestration is at the intersection of "eat" and "avoid being eaten," because through the herbivore's leafy diet, a sequestered toxin is used to avoid being eaten by predators.

If the plant produces toxins that are ultimately sequestered by herbivores for their own defense, then perhaps natural selection will favor reduced production of plant toxins. E. B. Poulton, a British evolutionary biologist, issued a call in 1914 for chemists to collaborate with biologists to solve the mystery of what makes monarch butterflies distasteful. Was it indeed the consumption of toxic substances from their milkweed host plants? And what are those substances? It took some sixty years to answer this question. Leading the charge was a group of talented scientists, including the Nobel Prize–winning Swiss chemist Tadeus Reichstein, the British naturalist Miriam Rothschild, and the young American lepidopterist Lincoln Brower. Not only did these scientists identify monarchs and milkweeds as one of the premier species pairs through which to study coevolution, but they also helped to establish chemical ecology as a discipline in its own right.

The story begins with Dame Miriam Rothschild, a naturalist, one of the founders of the field of chemical ecology, and someone who had a knack for bringing people together. She was an eccentric force of nature, known for her love of fleas, her "delightfully disheveled garden," and her flamboyant purple dresses, typically coupled with a kerchief around her hair and high-top sneakers. She had no formal education, and yet she had a keen sense of the biology of organisms and for frontier questions in science. Rothschild had been studying parasites and butterflies, among other things, and had been interested in their toxicity and mimicry. In the early 1960s, she encouraged a graduate student at Oxford, John Parsons, to pursue the toxicity of insects feeding on milkweeds. Their choice of a toxic food and their advertisement with bright colors had suggested to Rothschild, as it had to naturalists before her, that sequestration (the accumulation of toxins from the host plant) was likely.

Indeed, the toxic properties of milkweed were reminiscent of another plant, foxglove (*Digitalis*, from which the well-known drug takes its name), recognized to be emetic (vomit-inducing) for centuries. We now know that both milkweeds and foxglove contain compounds called cardenolides, and in the next chapter I will examine the essential chemical aspects of cardenolides in order to understand the milkweed-monarch arms race. Rothschild was interested in the hypothesis that similar chemical toxins in milkweed were finding their way into monarchs, making them distasteful and vomitous to bird predators. Although Parsons was not able to trace cardenolides moving from milkweeds to monarchs (could the monarchs be making them independently?), he did show the digitalis-like properties of several milkweed-feeding insects in a series of papers from 1963 to 1965. The last of these studies showed that monarch butterflies contained such a substance in the chrysalis and adults. Parsons purified the compounds and showed that they had toxic effects on frog hearts, guinea pig intestines, the blood pressure of cats, and enzymatic activity of human blood cells, and that they also caused starlings to vomit.

Next enter Tadeus Reichstein, a Swiss steroid chemist, fascinated by both animals and plants. In 1950, he was awarded the Nobel Prize in physiology or medicine (with E. C. Kendall and P. S. Hench), for work that resulted in the discovery of cortisone, one of the most important stress hormones in animals. Only one year later, in 1951, he described cardenolides from close relatives of milkweeds (the same class of steroidal compounds that give *Digitalis* its kick). And a decade later, in 1964, Rothschild wrote a letter to Reichstein, requesting his assistance in isolating cardenolides from monarchs. At the time, there were no known steroids derived from insects, and accordingly this was a fateful challenge for the renowned chemist. Rothschild and Reichstein engaged in a collaboration that would last more than a decade, and their first preliminary results were reported by Rothschild at a conference in 1966.

Lincoln Brower, a graduate student at Yale University, received his doctorate in 1957, working on the evolution of swallowtail butterflies. He quickly became obsessed with the hypothesis of mimicry among butterflies and the idea of sequestration, the notion that butterflies could accumulate toxic chemicals from host plants, thereby using these for their own benefit. In fact, his laboratory at Amherst College did much of the rearing for Parsons's, Reichstein's, and Rothschild's work, generating kilograms of monarchs (literally thousands of butterflies) for chemical analyses conducted in Europe. Not having access to the right equipment, and lacking abilities in chemistry himself, Brower and his colleagues took an unimaginably novel approach for their own studies of sequestration. Brower took thousands of monarch eggs and attempted to rear them on cabbage, which he assumed to be a benign host plant without toxins. If monarchs were gaining their toxicity from milkweeds, a cabbage-reared monarch would not be distasteful. His critical assay was not for the presence or absence of a chemical, but the behavior of a bird. In other words, Brower addressed the ecological consequences of the monarch's diet: would birds vomit if they ate monarch adults reared on milkweed (fig. 2.5), but not if they were reared on cabbage? He reasoned that if sequestration of plant poisons was important, birds would feed without nausea on cabbage-reared monarchs.

The reason this approach seems unimaginable is that, in nature, monarchs eat only milkweed. Period. If you asked me, after studying monarch caterpillars for more than a decade, could I ever get them to eat cabbage, my answer would be a resounding no, and without a second thought. So, how did he do it? Over five butterfly generations, Brower reared monarch caterpillars bit by bit,

FIGURE 2.5. Lincoln Brower's famous images of a blue jay barfing after feeding on a monarch butterfly. This highly repeatable assay of monarch toxicity usually concludes within twelve minutes.

only as long as they survived on cabbage. Hatched from thousands of eggs, nearly all the caterpillars were destined for a quick death by starvation, since they simply did not eat the unfamiliar food. In the first few generations, the larval survival rate was very low, with most of the caterpillars that did attempt to feed dying midway through development. Caterpillar death was not likely due to cabbage being deficient in some essential nutrient, but it could have been the result of cabbage's own defense compounds (mustard oils, or glucosinolates), or perhaps more likely because monarchs simply did not recognize cabbage as food. Whatever the case might have been, Brower persisted, switching them back to milkweed when they were near death. After this persistent set of rearings, in the fifth generation, Brower had nurtured a few caterpillars to adulthood that had fed only on cabbage and never on milkweed.

Working with wild-caught blue jays, and assaying their frequency of vomiting, he had an elegant experimental design. The birds were fed one of four foods: (1) monarch caterpillars reared on tropical milkweed (*Asclepias curassavica*), (2) monarch caterpillars reared on cabbage, (3) monarch caterpillars reared on a tropical milkweed vine called *Gonolobus*, or (4) mealworms, a most tasty food for birds. The milkweed-fed monarchs served as a positive control, scientific parlance for a treatment that should elicit an expected result: a barfing blue jay. Mealworms served as the negative control, since there was no expectation of vomiting. Brower's critical result was that cabbage-fed monarchs did not elicit vomiting. And furthermore, *Gonolobus* ended up being the exception that proved the rule. Brower had fully expected monarchs, which do feed on *Gonolobus* in nature, to be toxic when feeding on this milkweed vine. But, alas, the birds did not vomit. Much to his excitement, when Brower sent *Gonolobus* leaves to Reichstein in Switzerland, the results came back negative—no cardenolides. This was the confirmation that he needed to implicate plant toxins, which were now shown to be variable in host plants, as agents of the monarch's toxicity. Among Brower's four treatments, only when monarchs were fed a milkweed with cardenolides did they elicit a vomiting response from blue jays. The results were published in 1967.

In 1968, Reichstein, Parsons, and Rothschild published a study that was the last important step in nailing down proof of the monarch's sequestration. In fact, monarchs had two concentrated cardenolides, calactin and calotropin, in their adult bodies, and these same cardenolides were found in their milkweed host. The authors reasoned, "The fact that the cardioactive toxin of the monarch butterfly is of the cardenolide type . . . supports the suggestion that it is derived from the food plant and stored either unchanged or with only minor metabolic transformation." This paper, building on the three previous studies, sealed the deal on the paradigm of sequestration.

Later on, as the sequestration paradigm was cemented, some tension occasionally arose between Rothschild and Brower. They continued to work independently on monarchs and their sequestered toxins for decades, and perhaps both worried about their respective legacies. Who would be remembered as the discoverer of monarch sequestration? A few points about this scientific tension are worth explaining. First, it was Parsons and Reichstein (not Rothschild or Brower) who did the early heavy lifting in terms of physiology and chemistry, inspired and aided by Rothschild and Brower. They could not have done this work themselves. Second, Rothschild and Brower were themselves collaborative and reciprocally inspiring. In each of their respective early publications, the acknowledgments section is very telling. In most scientific publications, acknowledgments are provided at the end—usually crediting colleagues who contributed substantially and generously, but not enough to warrant being a coauthor of the study. The spirit of the acknowledgments in these four key studies is one of excitement, collaboration, and sharing. Letters, live butterflies, dried leaves, ideas, and inspiring words were being shipped across countries and oceans, with nearly all names appearing in all four papers' bylines or acknowledgments. That is what made this science move forward and, more generally, what makes science great fun. Despite the tension between them in later years, Rothschild and Brower share the legacy of being pioneers of chemical ecology and hugely important in the development of knowledge about monarchs and milkweeds.

THE DARWINIAN DEMON

The fact that monarchs sequester cardenolides from their milkweed host plants does not mean that these herbivores have won the arms race. As we will learn, the plants have responded and will continue to do so. Nonetheless, their coevolutionary escalation cannot go on forever. It does have limits. All evolutionary adaptations are constrained by a trade-off between energy investment in traits and growth or reproduction, a concept known to biologists as resource allocation. Investment in a beneficial behavior, physical feature, or internal physiology comes at a cost, because that energy cannot be used for some other function. The trade-off concept can be visualized as a Y-shaped tube, whereby all the energy that flows in from the bottom must flow to one of two (or a few) pathways. This Y-tube model has been very influential, probably because it is simple and intuitive. With a limited pool of resources, and nearly all adaptations taking energy, trade-offs must occur (fig. 2.6). Simply put, an organism cannot maximally grow, defend against predators, attract pollinators, and fend off competitors all at once.

Darwin and his predecessors recognized the importance of these counterweights in biology, as early crop and animal breeders realized that no organism could be bred to have all of the desirable traits. Indeed, because each trait is costly and trades off with allocation to other traits, breeders often develop specialized varieties that favor some traits at the expense of others. In the first chapter of *On the Origin of Species*, Darwin wrote: "As Goethe expressed it, 'in order to spend on one side, nature is forced to economise on the other side.' I think this holds true to a certain extent with our domestic productions: if nourishment flows to one part or organ in excess, it rarely flows, at least in excess, to another part; thus it is difficult to get a cow to give much milk and to fatten readily. The same varieties of the cabbage do not yield abundant and nutritious foliage and a copious supply of oil-bearing seeds."

Monarchs and milkweeds are subject to the same rules, and trade-offs are important in their evolution. We find high levels of genetic variation in toxin

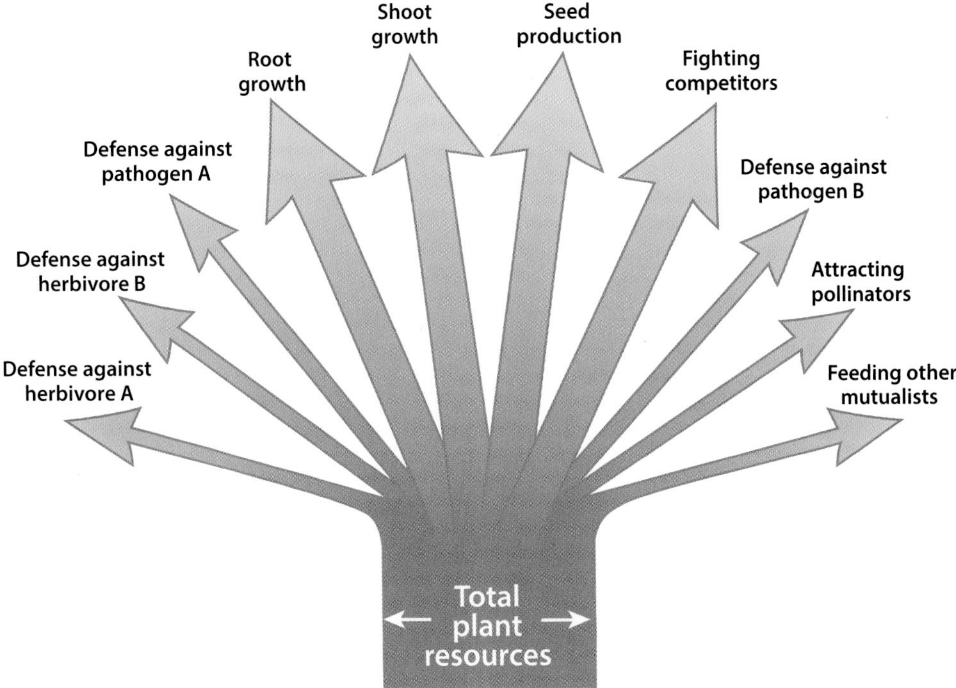

FIGURE 2.6. This diagram shows the allocation of limiting resources. Such diagrams are often drawn with simply two arrows, with energy flowing one way or another, say to growth or defense in a plant. It is now widely understood that even for a subset of the energy budget of an organism, there are many competing demands.

production among populations of common milkweed. The production of milkweed's major toxins (cardenolides) counterbalances with plant growth (fig. 2.7). Even though cardenolides are costly to make, under some conditions or in some years (for example, when soil nutrients are plentiful and when herbivore pressure is high), natural selection favors producing more. Nonetheless, under other conditions, the costs outweigh benefits, and plants are favored to produce minimal amounts of toxin.

Yet, in addition to the energetic drain caused by the production of a toxin (or any other trait), there may also be ecological offsets. Ecological trade-offs are experienced only under certain conditions and are not due to resource

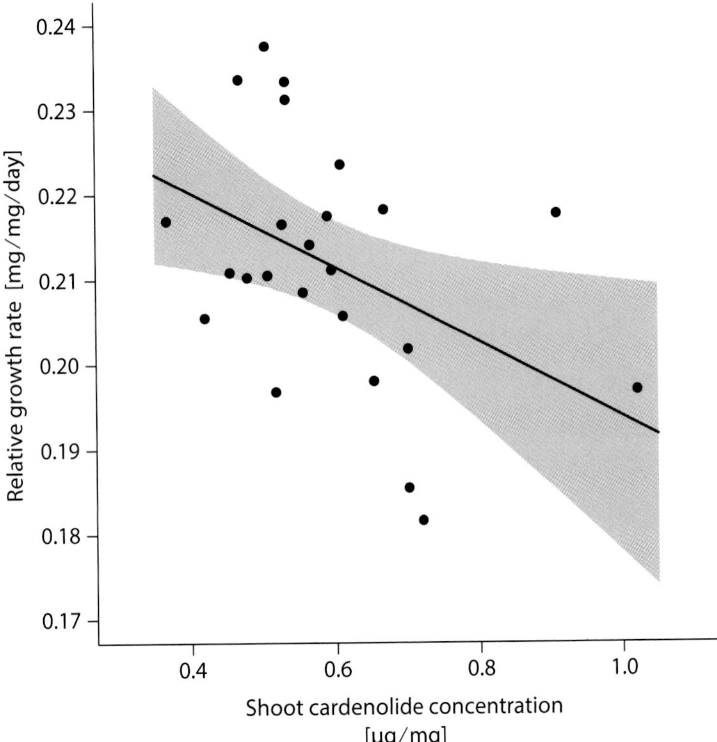

FIGURE 2.7. As common milkweed plants produce more cardenolides, they have slower growth rates. In this graph, each dot represents a genetic family of plants (each family having the same maternal and paternal parent plant), and the shading is the 95 percent confidence interval. Note the *x* axis, which shows that these genetic families exhibit two-fold variation in leaf cardenolide production. The families that produce more cardenolides have proportionally reduced growth rates, demonstrating a cost of producing defense chemicals.

limitation itself. Take, for example, a trait like the mating calls of crickets, which not only bear the cost of production, but also the ecological trade-off of attracting enemies. The chirps of a cricket are information released in the environment, and they may be used by any organism. Professor Marlene Zuk, from the University of Minnesota, found that cricket songs used in mating were also specifically attractive to a parasitic fly. This is just the sort of trade-off that can affect the evolution of an organism's traits, and perhaps even coevolution. Ecological trade-offs work alongside energetic costs and represent conflicts of interest.

In coevolutionary interactions, when one species is imposing natural selection on the other, and the second species is adapting and reciprocally imposing selection on the first, trade-offs can limit the speed and intensity of an arms race. One can just imagine the arms race grinding to a halt, or ending in stalemate as it were, as the respective species' armament is simply becoming too burdensome, weighty, or expensive. In other words, species may reach a tipping point in either the energetic expenditure or ecological cost for some particular defense that has too high a price for too little benefit. In the case of monarchs and milkweeds, there are many trade-offs on both sides, with not only energetic drains, but also ecological costs associated with offensive and defensive traits.

The discovery that milkweeds do not need monarchs for pollination, but that monarchs need milkweed, both for food and defense, helps us to understand evolution and coevolution through the lens of trade-offs. Although the cardenolides of milkweed provide defense against insects, they also provide protection to one of its major pests, the monarch butterfly. As the milkweed has evolved more toxic cardenolides, monarchs themselves face an ecological trade-off, that of the negative effects of ingesting cardenolides and the positive protective effects of sequestering them as a defense against their own predators. The balancing of such conflicting interests represents a general dilemma faced by all organisms, one that ultimately limits the extent of adaptation. It simply may not be possible for an organism to do it all—and certainly not all at once.

Because of trade-offs, most species tend to harbor a substantial amount of genetically based variation in traits. Under some conditions, for example, when resources are plentiful or when a major predator is present, the most costly defensive trait may be favored. Under other ecological conditions, however, other traits will be bolstered. Because the combinations of environmental conditions are endless, and conditions vary across space and time, most species live as multiple populations, in a patchwork being pushed and pulled by natural selection in different ways. This is how and why species maintain a lot of genetically based variation in traits. Think of this view of evolution as a giant pinball machine, with the ball being flung in all sorts of directions across the landscape. The ball represents the traits of a species and is flung in different directions by the forces of natural selection, which vary in space and time. Over the long haul, when we zoom out, the result is the maintenance of variation within a species. As long as the far-flung individuals are at least occasionally favored by natural selection, and are also exchanging genes with (mating with) other individuals of the species, the species will continue to harbor variation. Such genetic variation is the raw material for natural selection to work on. And when species are engaged in an intense arms race, spectacular outcomes are possible.

Coevolutionary warfare between monarchs and milkweeds has made use of extraordinary chemical compounds. The cardenolides introduced in this chapter are at the center of their struggle. Although the chemistry of milkweed defense and monarch sequestration is not unique, it is extreme, and it serves as an important model to generally understand how chemical compounds can be a focal point in coevolution. The persistence of their ecological and evolutionary interaction has led to plants that produce heart-stopping toxins and butterflies that have derived molecular solutions to this poisonous challenge.

CHAPTER 3

The Chemistry of Medicine and Poison

> Ut quod ali cibus est aliis fuat acre venenum.
> (What is food to one, is to others serious poison.)
>
> —Lucretius, *De rerum natura* (*On the Nature of Things*), book 4

Milkweeds are continually evolving to be better-defended fortresses, and monarch butterflies are powerful enemies developing strategies to break down the plants' barriers. Looking to milkweeds as defenders and monarchs as exploiters not only provides a tale of natural history and lessons in reciprocal adaptation but shows how chemistry became one of the focal points in a coevolutionary war. For a caterpillar feeding on milkweed, high levels of toxins in the leaf serve both as poison and also as a medicine that the caterpillar will sequester away. As I noted in chapter 1, the axiom of the Renaissance scientist Paracelsus, "dose makes the poison" certainly applies to the toxic principles of milkweed. And the tipping point between medicine and poison is one of the most critical issues in the study of coevolutionary arms races.

Our understanding of the role of milkweed toxins in interactions with monarchs comes, in part, from an examination of the use and abuse of these compounds by humans from around the world. Indeed, as we will learn later in this chapter, the effects of milkweed toxins are very general, and essentially all animal cells react in the same way when exposed to them. But once again, dose is everything. In his classic 1892 book, *Medicinal Plants*, Charles Millspaugh cataloged the medical uses of milkweed and their mighty array of effects on the human body as a "subtonic, diaphoretic, alterative, expectorant, diuretic, laxative, escharotic, carminative, anti-spasmodic, anti-pleuritic, stomachic, astringent, anti-rheumatic, anti-syphilitic, and what not." Native Americans, from the Cherokee to the Chippewa, used various milkweed species, including common milkweed, as a medicine for backaches and warts, as a contraceptive, and to treat congestive heart failure. As a poison, milkweed extracts are used in the majority of all African poison arrows employed in hunting and warfare. That dose makes the medicine (and poison) was not lost on the Romans, who also used botanical extracts containing milkweed toxins as a heart tonic as well as a rat poison.

Like all plants, milkweed contains a soup of many chemical properties. Nonetheless, the lion's share of the attention paid to milkweed has been focused on a single class of chemical compounds introduced in the previous chapter: the cardenolides. Some of the most intriguing interactions between monarchs and milkweeds involve this group of steroidal compounds, which are made up of carbon, hydrogen, and oxygen.

Before we really dive in, however, I have a confession. I have always been somewhat of a chemophobe. Not afraid of chemicals per se, but of the study of chemistry. The lore around the difficulty of chemistry courses at universities is legendary, especially organic chemistry (or "orgo" as it is frequently called), the study of chemical compounds that contain carbon. When I was a student, and now as a professor, I have frequently heard horror stories about organic chemistry. And yet, most college curricula in biology require several semesters in chemistry, including orgo. And it makes sense, because carbon is the essential backbone of all living organisms. Yet, resourceful students typically find a way to avoid chemistry courses—and so, too, did I find a loophole. At the University of Pennsylvania, I carefully chose a concentration in biology that did not require taking organic chemistry.

Having chosen to specifically avoid taking these courses, I have been paying the price ever since. It is somehow ironic that I fell in love with two organisms that are inextricably linked by chemistry. They have spent millions of years evolving chemical traits and reciprocally coevolving in a manner that puts chemistry at the center of their arms race. Now, each spring when I teach in our chemical ecology course at Cornell, my co-instructors are always amused by the last-minute elementary questions I ask them before lecturing, just to make sure I truly understand the little bit of orgo I need to explain. Alas, what is done is done, and much of the work of professional scientists is, indeed, teaching ourselves what we need to know to get the job done. My hope is that, given my lack of formal training in organic chemistry, this chapter will be understandable to any reader.

YELLOW VISION

The medicinal properties of cardenolides have been known for a few hundred years. And some of the ailments that they treat have long been known, while others are still being discovered.

Consider the genius and tragedy of Vincent van Gogh. In the last two years of van Gogh's life, his paintings took a turn toward yellow and glowing colors. Think of his starry night, wheat fields, and sunflowers. Thomas Lee, in a 1981 article in the *Journal of the American Medical Association*, noted that for the two years before his death, van Gogh was likely prescribed a tincture containing cardenolides by his doctor, Paul-Ferdinand Gachet. The foxglove plant (scientifically known as *Digitalis*), although not a milkweed, contains cardenolides, the very same class of chemical compounds found in milkweeds. And although this is speculation, Doctor Gachet likely prescribed foxglove tea to van Gogh. The flowers appear in the foreground of his two paintings of the doctor (fig. 3.1). In the second half of the nineteenth century, *Digitalis* was prescribed for epilepsy, one of the many conditions that plagued van Gogh. And we now know that two side effects of too high a medical dose of cardenolides include yellow vision and seeing halos around bright objects such as lights. It is certainly possible, and apparently likely, that van Gogh's shift in color pallet and style of painting in the last two years of his life were due to cardenolide intoxication. Whether or not this helped with his many ailments, however, is unclear.

FIGURE 3.1. The cover of the February 20, 1981, issue of the *Journal of the American Medical Association*. Study of Van Gogh's famous painting of Dr. Gachet led to a medical hypothesis (published in this journal) about how plant toxins affected his artistic style.

About one hundred years before van Gogh's passing, William Withering, an English physician who was also a botanist, chemist, and geologist, discovered the medicinal properties of cardenolides from foxglove. He is said to have learned of the benefits of foxglove for treating heart ailments from a gypsy herbalist. At the time, foxglove was also used for other medical treatments as well, some in trial and error, probably like the treatment of van Gogh. And remarkably, there are reports of the use of other botanicals with the same chemical compounds for heart problems and other ailments dating back to ancient Egypt and Rome.

Withering investigated the beneficial effects of foxglove for patients with "dropsy," what is now known as pulmonary edema, or, more simply, the swelling of body tissues (often the legs) owing to the inability of the heart to keep up with the body's need for blood (also known as congestive heart failure). In 1785, he published a summary of medical cases in which foxglove was administered to patients with sundry illnesses, from epilepsy to insanity, but largely also with symptoms of dropsy. In the more than one hundred cases studied, he documented the clinical effects of extracts of foxglove, many of which improved health. He also summarized knowledge of foxglove extracts from other doctors, including Erasmus Darwin, grandfather of Charles Darwin, similarly a physician and botanist, who was also a poet and early developer of ideas on evolution. Erasmus Darwin was also sold on the benefits of foxglove in the treatment of symptoms relating to dropsy. After ten years of trials and investigation, Withering concluded in his book that foxglove was a strong diuretic, effective in different forms of dropsy, potentially useful in a diversity of diseases, and, most profoundly, "That it has a power over the motion of the heart, to a degree yet unobserved in any other medicine."

Because of the nearly identical chemistry of foxglove and the milkweeds, our story is not only richer, but also more complete. The bulk of plants that make cardenolides come from a single botanical family named the Apocynaceae, "the dogbane family," which includes the milkweeds among some 1,500 other species. Cardenolides also appear sporadically in a handful of other botanical families. Interestingly, foxglove resides in a distantly related family, the Plantaginaceae. These facts illustrate two conceptually important issues in evolutionary biology. First, some traits are evolutionarily conserved, that is, most related species share this set of conserved traits. Just as hair is a conserved trait among mammals, so too are cardenolides conserved in the botanical family that includes milkweeds (the Apocynaceae): most species have cardenolides because they inherited this trait from their ancestor. Second, when the same trait independently evolves in distantly related species, the trait is said to exhibit "evolutionary convergence." Just as flight has independently evolved in birds, mammals (bats), and insects, so too have cardenolides independently evolved in several plant families (fig. 3.2). The same trait, here cardenolides,

FIGURE 3.2. (a) The common milkweed (*Asclepias syriaca*) and (b) common foxglove (*Digitalis purpurea*), two distantly related plants that convergently evolved to produce cardenolides.

can both be conserved, because it is maintained among many closely related species, and convergent, because independent innovations of the trait occurred at least a few times in unrelated plant families. In fact, although foxglove is famous for its medicinal cardenolides, we now know that the diversity of cardenolides present in the milkweeds is likely far greater. The chemicals have convergently evolved in these two groups, but they diversified to a greater extent in the milkweeds.

To this day, cardenolides are commonly prescribed in the treatment of congestive heart failure. They are currently also burgeoning as therapeutics for other medicinal needs, the most promising of which is as an anticancer drug. Cardenolides were recently found to improve the survival of cancer patients, leading to current clinical trials. They have also been recognized as naturally occurring human hormones, which induce a cascade of cellular and physiological effects. Cardenolides have antimicrobial effects, which would not have been predicted given the chemical structure of the compounds. Countless patients owe their lives to this wonder chemical from plants, and it appears that this is only the tip of the iceberg.

CHEMISTRY 101

What exactly is a cardenolide? At its core, a cardenolide is a steroid, and steroids are characterized by four chemical rings, totaling seventeen carbon atoms. Steroids are common and well known in living organisms, including compounds such as testosterone and cortisone. But what makes a cardenolide a cardenolide is that another ring, a "lactone," is attached to the steroid. Lactones too are carbon rings, but one of their carbon atoms is replaced by an oxygen atom, and they have another oxygen atom attached to the end of the ring by a double bond. To complete the picture, cardenolides contain what are termed "side chains," which can be variable chemical structures depending on the specific compound (fig. 3.3). Cardenolides often occur as "glycosides," because a sugar molecule is attached to the steroid as well. These sugars can be highly variable in chemical composition and typically range from one to four sugar molecules bound to the steroid. If this terminology is too much detail for you, please don't despair; you can mostly ignore these specifics.

Cardenolides have evolved independently in more than twelve plant families, but as mentioned above, the majority of cardenolide-containing plants are in the Apocynaceae. In addition to the milkweeds, familiar plants in this family include dogbane, oleander, frangipani, periwinkle, and swallow-wort. Similarly, closely related defense compounds, termed bufadienolides, are produced in toads (in the genus *Bufo*) and also in a few plants. Hundreds of these compounds, produced by different milkweeds, by plants in other families, and by toads, all have the same "mode of action" (or effect) on animal cells. Mode of action is the mechanistic basis for how a chemical compound interacts with a target biological organism—meaning the animal's taste buds, cell membranes, gut, or other tissues. When you drink wine or eat an unripe piece of fruit, that

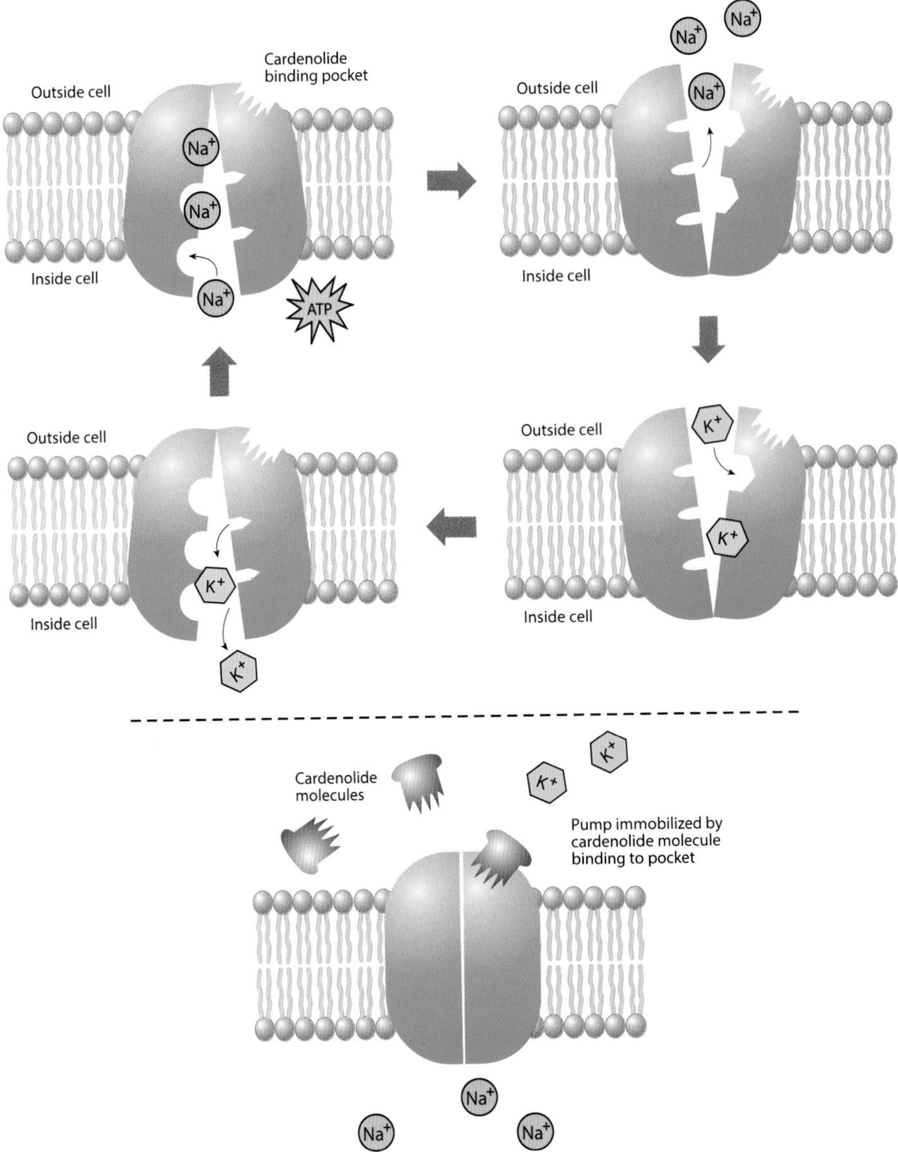

FIGURE 3.4. A conceptualization of the cellular enzyme, the sodium-potassium pump, which is common to all animal cells. For the price of one molecule of ATP, the pump removes three sodium ions (Na^+) from within the cell and shuttles in two potassium ions (K^+). In this conceptualization, there is a change in the binding affinity for the ions (note the change in shape) after release. When cardenolides attach to the binding pocket, the sodium pump becomes inactive, and cellular function is disrupted.

disproportional binding to the heart muscles, fifteen to thirty times higher than to blood itself, and two times higher binding to heart muscles than to skeletal muscles. Given the diversity of sodium pumps in the vertebrate body (four types), it is the superior binding of cardenolides to cardiac muscles that suggests why this chemical has such a direct effect on the heart. As a medicine, the right dose of cardenolides causes stronger (but slower) heart muscle contractions, which can increase blood flow per heartbeat when a person is experiencing congestive heart failure.

Now, why do the ubiquitous and ever-important sodium pumps have pockets to which a chemical compound that disrupts its enzymatic function can bind? Although definitive evidence is lacking, the latest physiological evidence is promising as an explanation. Beginning in the early 1980s, it was suspected that vertebrates (including humans) used a self-produced hormone to regulate heart rate and strength of contraction. Technically, a hormone is any chemical compound produced by an organism (plant, animal, or microbe) that regulates some physiological or behavioral process within an individual. After a decade of searching, researchers discovered that humans produce ouabain—most likely the same exact cardenolide produced by some plants. Continued biomedical research aims to identify the self-produced hormonal regulation of cardiac function. But for our purpose, we can simply understand that all animals possess sodium pumps, and some animal tissues appear to have sodium pumps that are more sensitive to cardenolide binding than other tissues. For insects, the majority of species appear to have a single form of the sodium pump that is produced in many different tissues (although this has not been well-studied), but especially that of the nervous system. It is currently unknown whether insects themselves produce cardenolides as hormones to regulate bodily functions. I suspect not. But then again, just a few decades ago it would have been unfathomable that humans produced their own cardenolides as a hormone.

Regardless of why animals have susceptible binding pockets, it is relatively clear why plants and some toads produce these steroidal toxins. They make them as a defense. But as with all toxins, "dose makes the poison," and in an arms race, natural selection drives the evolution of higher doses. These toxins evolved in coevolution by killing herbivores that were eating plant tissues. The effects as a heart tonic are surely incidental. And although it is unclear whether the primary evolutionary target of the toxin was the vertebrate cardiac form of the sodium pump, or simply the more general sodium pump expressed in insects, cardenolides have no known function for the plant other than defense. Just as a dash of ground cayenne chili may be a flavoring on pizza or poison in pepper spray, so too are cardenolides heart tonics and poisons. But the evolutionary driver of cardenolide evolution has been the plant's arms race with insect pests, including the monarch.

LIKE OIL AND WATER

As Dr. Withering was coming to conclusions about the medicinal value of cardenolides in the late eighteenth century, he noted that the optimal dose to reduce the symptoms of heart failure is achieved when patients just start to have indications of "intoxication," namely, vomiting and yellow vision. As the various botanicals, extracts, and tinctures were prepared, it became clear that some preparations of cardenolides may be much more active and powerful than others. The various preparations involved using different plant species, extractions in alcohol versus water, and orally ingested extracts versus those that were injected directly into the bloodstream (although injections did not become common until the mid- to late eighteenth century). Thus, in addition to "dose," the specific identity of a cardenolide, determined by its side chains and sugars (see fig. 3.3), is critical to its biological impact.

The ease with which a cardenolide enters an animal's bloodstream is of particular importance to understanding how these compounds affect that animal's metabolism; some readily enter the bloodstream, while others do not. The mouth of any animal is one of the few open windows to the body. Yet, as so many different substances enter the mouth, the animal "gut" or tube between the mouth and the anus, is a massive barrier to substances entering the bloodstream. Simply put, substances that are not absorbed by the time they reach the end of the small intestines will pass out the back door without entering the bloodstream. Some cardenolides that are ingested never find their way into the bloodstream because the gut is a strong barrier, and the chemicals do not diffuse in. Even substances that are absorbed may be selectively shunted to organs to be metabolized or discarded.

To understand why some cardenolides are more toxic than others, we must introduce the concept of chemical polarity. Although at its core, polarity is determined by electrical charges of compounds, which have consequences for chemical bonding, what is most critical to understand is that cardenolides come in forms from highly polar to highly nonpolar, mostly determined by the chemical composition of the attached sugar molecules. Because water is a highly polar compound, and as all chemists teach, "likes dissolve likes" highly polar cardenolides are typically soluble in water. We also know from the chemistry of everyday life that oil and water don't mix. Alcohol is somewhere in between, less polar than water, but more polar than oil. Nonpolar substances like oil typically cannot dissolve in water and only mix with "likes." This is why soap is effective for washing our hands and clothes, as it has nonpolar (hydrocarbons) and polar (salts) components. The soap forms pockets that trap the nonpolar substances, but those same pockets are water soluble, so they wash away.

Differences in the chemical polarity of cardenolides can make all the difference in their medicinal or toxic properties. Polar cardenolides are most likely to be passed out by an animal body (be it human or monarch butterfly), while nonpolar cardenolides are likely to be absorbed into the bloodstream. Over the past hundred years or so, two cardenolides, digitoxin and digoxin, have often been prescribed for heart patients. These two compounds, however similar sounding in name, are not equivalent. Digitoxin is highly nonpolar and requires a smaller dose to be effective, while digoxin is more polar, and requires a stronger dose to have an equivalent effect. The appropriate medicinal dose is in part determined by how well the cardenolides are taken up and can move throughout the body. Because digitoxin is a nonpolar compound, it can easily cross the lipid (fatty) membranes that surround cells, enter our bodies, and get to work. Meanwhile, the more polar digoxin is not easily absorbed by the human body, and is readily discarded by our kidneys.

In 1968, a major pharmaceutical blunder demonstrated the above point about polarity and differential effects of digitoxin and digoxin. Many patients were taking a 0.07 mg daily dose of digitoxin for their heart conditions. The equivalent medicinal effect could be achieved by a higher dose of digoxin, usually about fourfold higher, or 0.25 mg per day. The mix-up occurred in the town of Veenendaal, Holland, where 179 heart patients, the bulk of whom were over sixty years of age, inadvertently took an overdose of cardenolides for several weeks, when tablets were produced with 0.20 mg of digitoxin (and 0.05 mg digoxin), but were labeled as pure digoxin. In other words, patients were given what would have been the correct dose of the less potent digoxin, but were instead overdosed with digitoxin. In this tragic case, six patients perished, and more than 90 percent suffered symptoms including yellow vision, vomiting, and extreme fatigue. As a medicine, cardenolides are miraculous. And as a toxin, cardenolides are occasionally still reported to cause human mortality, usually by accident or in homicide or suicide.

Perhaps the human body and that of a monarch butterfly seem radically different, yet some chemical properties impact the two species in the same way. Research in my laboratory has also found that the nonpolar digitoxin is among the most active cardenolides affecting monarch caterpillars. Understanding the chemical attributes that make certain cardenolides more or less toxic is at the heart of modern questions about coevolutionary arms races.

COEVOLUTIONARY CHEMISTRY

A fellow butterfly biologist once described to me a caterpillar feeding on a plant that was so toxic that the caterpillar was "choking." This is one of those

FIGURE 3.5. A monarch caterpillar that died two days after starting to feed on *Asclepias linaria* from the Sonoran Desert in Arizona.

anthropomorphisms that is hard to understand until you witness it. When monarch butterflies, with all of their adaptations to milkweed, feed on some of the most toxic species, it appears there is little way for them to fight through. They basically choke. I once observed this in my greenhouse, when I grew monarchs on *Asclepias linaria* from Arizona, one of the highest cardenolide-producing milkweed species (fig. 3.5). In this case, most of the caterpillars took some bites, ate a bit, and perished. Those that survived grew slowly and were sickly. Sometimes on such highly toxic milkweeds, monarchs will go after the flowers—perhaps they are lower in cardenolides? But there is essentially no escape.

Cardenolides are found in virtually all of the plant parts that have been examined, including roots, shoots, and fruits. But cardenolides are also abundant in the foamy pith tissue inside of stems, in the phloem sap (or sugary liquid that flows through plant plumbing), flowers, and even floral nectar. In chapter 7, which examines the community of insects attacking every conceivable plant part of milkweed, I will return to this issue of the distribution of cardenolides

FIGURE 3.6. The larvae of the common crow butterfly (*Euploea core*), an early diverging milkweed butterfly that consumes cardenolide-containing plants, do not take up cardenolides into their bloodstream (they have a sensitive sodium pump). A healthy caterpillar is shown on top. When cardenolides are injected into their bodies, caterpillars stop feeding and become flaccid, not holding on to leaves. After twenty-four hours, most do not recover. When monarch larvae are injected with the same dose of cardenolides, no effect is observed (they have a largely insensitive sodium pump).

among the parts. For now, suffice it to say that the concentrations are often high, making for a bitter, vomitous, and toxic food plant.

When cardenolides are experimentally given to insects, most experience negative effects. For example, when gypsy moth and silk moth caterpillars are fed diets laced with cardenolides, they are poisoned. Many mammalian and insect herbivores are so strongly deterred from feeding on foods that contain cardenolides that assessing toxicity when ingested has been difficult. Thus, taste aversion allows for avoiding the toxic properties of milkweeds. Studies employing cardenolide injections have shown the toxic effects of cardenolides by circumventing taste aversion and any barriers to uptake of cardenolides in the gut (fig. 3.6).

And yet cardenolide-containing milkweeds are the only food for monarchs. In fact, monarchs frequently sequester well over 0.25 mg of cardenolides in their bodies, the daily dose given to treat heart patients. Just think about that: an adult butterfly weighs less than one gram, and an average adult human weighs about 150 pounds (more than 68,000 grams). The butterflies are tolerating an extraordinarily high concentration of cardenolides, thousands of times more cardenolides per gram of body mass than humans could tolerate. How do they do it? For years it was a long-standing mystery, but in the late 1970s, the story of monarchs' relative insensitivity to cardenolides began to unravel. A physiological study at that time showed that the sodium pump, when extracted from monarch brains, was three hundred times less sensitive to a standard cardenolide (ouabain) than those of other Lepidoptera that do not typically eat milkweeds (fig. 3.7). This was a major discovery in the field of insect physiology, but the study did not initially have a strong impact on research into the ecology and evolution of monarchs. Because information in some subdisciplines of science is often not well-communicated to other subdisciplines, it took some decades before additional strides were made.

What gives monarchs their relative insensitivity to cardenolides was discovered in the 1990s, when a graduate student, Ferdinand Holzinger, studied with the renowned German chemical ecologist Michael Wink. Together they discovered that monarchs harbor a single DNA mutation in the gene that codes for the sodium pump, and this mutation makes monarchs relatively insensitive to cardenolides. All animals require functional sodium pumps, and this enzyme originated early in the evolution of animal life. Nonetheless, it has not changed much since the origin of animals (the gene is highly conserved). Most animals, from puny worms to pigs, have the same sodium pump, which is sensitive to cardenolides. Only monarchs and other cardenolide-eating insects have modified their sodium pump.

Holzinger and Wink not only found a seemingly important difference in the gene sequences between monarchs and the several other insects they examined, but they went a step further and demonstrated the function of this

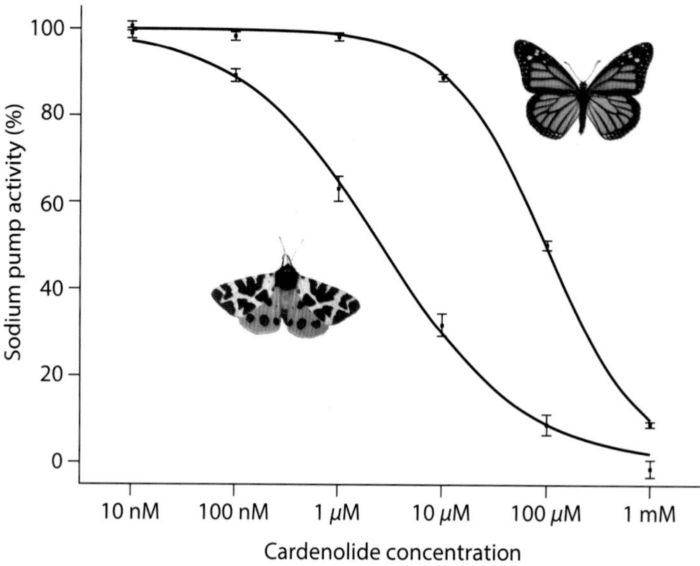

FIGURE 3.7. After extracting the sodium pump from the brains of the garden tiger moth (*left*) and the monarch butterfly (*right*), a postdoctoral fellow in my laboratory, Georg Petschenka, tested them for sensitivity to cardenolides in the laboratory. The standard cardenolide ouabain was dosed on the two, showing that the monarch is much less sensitive to cardenolides.

change. Genes are composed of long sequences of "bases," named A, T, G, or C (shortened forms of their chemical names). Each group of three bases is translated into an amino acid, and chains of amino acids form proteins. It is such proteins that make up the bulk of the machinery that conducts the functions of life. The sodium pump's proteins, for example, compose an enzyme of some 1,500 amino acids. The pocket (or alpha subunit as it is called), where cardenolides bind, is itself composed of about 750 amino acids. Astoundingly, Holzinger and Wink found that the amino acid at position 122 of the binding pocket was different in monarchs from the one in all other noncardenolide-feeding insects they sequenced, as well as that from pigs and humans. This evolutionary change (from asparagine to histidine at position 122 in the pump) was brought about by a single base-change, one of the three that make up the amino acid, and this sufficiently alters the physical shape of the sodium pump pocket so as to substantially reduce binding of cardenolides.

A DIVERSITY OF TOXINS

Plants possess an astounding diversity of chemical flavors, vitamins, medicines, and poisons. The cardenolides are but one tiny group of these, and they are potent toxins at the center of a specific coevolutionary arms race. Monarchs have cracked the code of cardenolides by mutations in their sodium pumps. However, perpetual insensitivity to a plant defense, at least over time, is not an option in coevolution. Once a defensive trait is rendered ineffective by the evolution of a counterploy in the herbivore, a plant must evolve a modified defense, lose this useless defense, or stack on additional defenses that make the original defense more functional. Keeping pace with each other is the name of the game.

Among the many plant defense traits favored by natural selection, cardenolides are generally effective against many herbivores. Such traits offer economy in the battle against diverse enemies, as one size fits all (or at least many). Nonetheless, in most coevolutionary arms races, one means by which plants cope with adapting pests is to diversify the specific types of toxins produced. Most milkweeds produce a multitude of distinct cardenolides. Different cardenolides have specific properties in terms of their ability to cross animal cell membranes, to bind to the animal sodium pump, and to cause toxicity. The typical milkweed plant produces more than ten different types of cardenolides in its leaves, from the polar to the nonpolar.

Several hypotheses have been put forward for why a single plant may produce different types of any secondary compound. The study of all sorts of drugs has taught us that mixtures or "cocktails" are often more effective than single shots of a chemical. This has certainly been true in terms of medicinal success in managing bacterial diseases, HIV, and cancer. Insecticides in agriculture are no exception; mixtures of compounds are used both to maximize the kill and to reduce the risk of evolution of resistance in the pests. Hence, there may be many benefits of milkweeds producing a mixture of cardenolides. Mixtures themselves may be more potent than an equal concentration of a single compound because the mixture can travel to diverse places in the body and may be difficult to compartmentalize or metabolize. Alternatively, mixtures may not be cocktails pointed at single pests, but the different chemical forms may target different pest species, with some cardenolides most effective against monarchs, but others more effective against milkweed aphids, or beetles, or even microbial pests. These are compelling hypotheses waiting to be tested.

Having mastered the basics of coevolutionary chemistry and having learned how cardenolides are potent toxins at the center of an arms race, we must now turn in a different direction. Our next stop on this journey takes us back to the beginning—to the beginning of each year, when an ecological cycle starts at the monarchs' overwintering grounds in Mexico. In this phase, the butterflies wait, mate, and then migrate, while milkweeds are dormant in the frozen north.

CHAPTER 4

Waiting, Mating, and Migrating

The woods are lovely, dark and deep,
But I have promises to keep,
And miles to go before I sleep,
And miles to go before I sleep.

—Robert Frost, "Stopping by Woods on a Snowy Evening"

Every January, when monarch breeding grounds in the northeastern and midwestern United States are covered in snow, millions of monarch butterflies huddle together in a few aggregations in the high mountains of central Mexico (in the states of Michoacán and México). During these winter months, monarchs are typically found on mountain peaks at ten thousand feet (three thousand meters) above sea level, where temperatures hover around ten degrees above freezing. The butterflies quietly bide their time and occasionally flutter in the sun. When the weather warms, they fly to drink water from streams, as precious nectar is scarce, owing to the lack of blooming flowers at this time and the high elevation.

How they got to Mexico is the subject of chapter 8. But having endured a journey of thousands of miles, they now perch in trees for four months of waiting in the cool mountain air. The butterflies are waiting for spring. As the days get longer and warmer, monarchs begin to mate. Eventually, they leave Mexico for their northern sojourn (fig. 4.1). Meanwhile, across the border in Texas, milkweeds are emerging from dormancy, growing from the roots of plants that senesced in the previous winter. These plants will provide a natal home for the new year's first generation of monarchs.

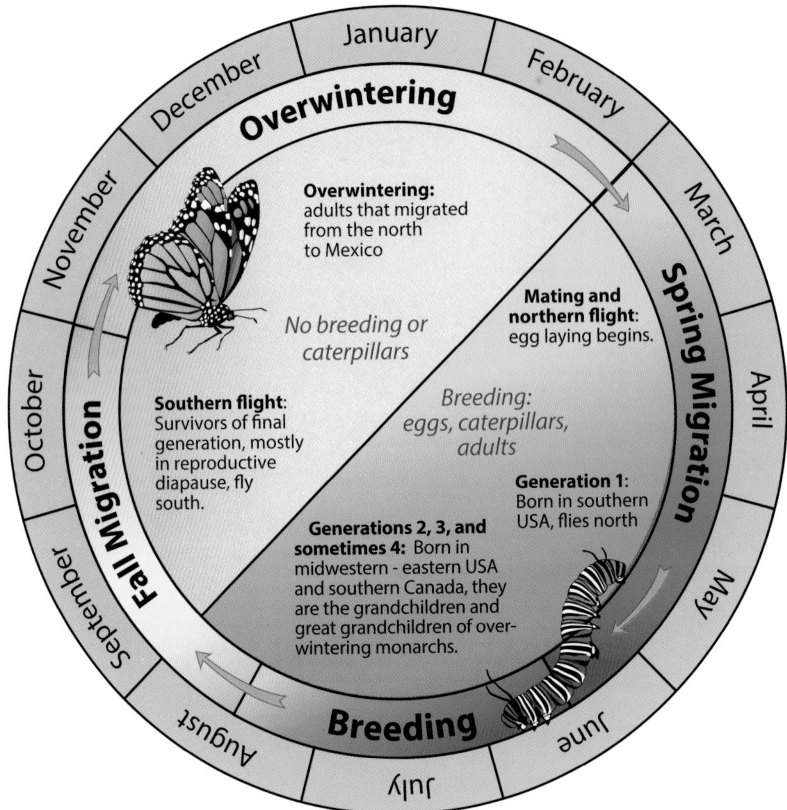

FIGURE 4.1. A diagram showing the annual migratory cycle of the monarch butterfly across eastern North America. Note that monarchs rely on milkweed only during the breeding season (March through August). Nonetheless, some breeding and a few caterpillars may be found year-round in the southern United States and Mexico.

During the first few months of each year, monarchs leave Mexico and migrate north to mate and lay eggs. How do the butterflies know that it is time to leave Mexico? What are their mating rituals? How far north do they travel during the spring migration before laying eggs? And how do females make decisions about where to lay eggs? Although the first two of these questions do not involve monarchs' interactions with milkweed, I will return to these plants and their arms race with monarchs to answer the latter questions.

MATING IN THE NEW YEAR

Picture this: a snowy New Year's Day in central New York, perhaps in Ithaca, where I live and am writing this coevolutionary odyssey. The average January temperature stays a bracing ten degrees below freezing. Nearly all plant and animal life is dormant. Now travel, as I did just before New Year's Eve in 2011, to Mexico City, some two thousand miles (3,200 kilometers) southwest as a crow flies. The climate in Mexico's capital, at seven thousand feet (2,150 meters) above sea level, is subtropical and hovers just under 65 degrees Fahrenheit (20 degrees Celsius) throughout the year. It is only a short drive from the city to the Monarch Butterfly Biosphere Reserve, primarily situated in the forested highlands of Michoacán State. Here, on about a dozen mountaintops, millions of monarchs spend the winter in a condition of reproductive diapause (fig. 4.2). As the word "dia-pause" suggests, the monarchs are on reproductive hold, not having fully formed their reproductive organs. Although their diet mostly consists of water from streams and condensed droplets on vegetation, as the days lengthen and the season warms, changes in monarch physiology begin.

At the overwintering sites, three thousand feet (one thousand meters) higher than Mexico City, winter temperatures fluctuate between 41 and 54 degrees (5 and 12 degrees C). The butterflies are typically quiescent. In order to fly, monarchs need air temperatures of about 55 degrees (13 degrees C), which is regularly achieved, if only briefly, in the winter sunshine. Then beginning in late January each year, a mating frenzy begins. By late February, as the temperatures regularly warm to above 60 degrees (15 degrees C), and the amount of daylight increases to nearly twelve hours, the monarchs become reproductively mature and ready themselves for the spring migration. As I will discuss in chapter 9, the narrow band of suitable temperatures for monarchs at the overwintering sites will likely be severely impacted by climate change.

Cold winter temperatures followed by a warming are critical for the switch in migratory behavior from flying south in the autumn to flying north in spring.

FIGURE 4.2. Overwintering monarchs sunning themselves at the Monarch Butterfly Biosphere Reserve in Sierra Chincua, Mexico. On warm, sunny days, butterflies temporarily leave their dense and tightly packed aggregations to fly in the sunshine and forage for water along the edges of streams.

A recent discovery by Patrick Guerra and Steven Reppert, of the University of Massachusetts Medical School, showed that a prolonged exposure to cool temperatures—around 46 degrees (8 degrees C)—is critical for monarchs to reverse their migratory compass from moving south to moving north (fig. 4.3). Quite remarkably, it appears that without the switch to cooler overwintering temperatures, monarchs would simply continue on their journey south in the spring! But before monarchs move north, mating must commence, and this process is not all about peace and love. Miriam Rothschild, introduced in chapter 2 as the heroine who advanced the discovery of monarch sequestration, described the affair as a "brutal onslaught." She went on to say, "The Monarch butterfly could well be designated nature's prime example of the male chauvinistic pig. . . . [He] more often than not, knocks down his female, and while in a half dazed state, takes her by force. Her antennae may be bent double in the process, her legs crumpled beneath her body and her wings sadly mutilated."

Monarch mating typically has two phases, one in the air and one on the ground. Both are rough, and neither appears romantic. Courtship typically begins with the male pursuing the female in the air. Occasionally, this aerial premating ritual involves something that looks quite similar to that of many butterflies, where the male and female fly in circles or spirals, with one following the other, and with the two eventually perching on a tree branch. However, the more typical scenario for monarchs is different. Although "courtship" is the term typically used, monarchs are a bit unusual in that a midair pursuit is often followed by escape behavior in the female, where she very quickly zig-zags away. Thomas Pliske, now retired from Florida International University, has described this fast and erratic female escape behavior as similar to that which occurs when bird predators pursue monarchs as prey. Nonetheless, if a male successfully grasps a female with his legs around her head and body and over and under her wings, the two drop to the ground, something Pliske termed "pounce and takedown" behavior. In only about 10 to 30 percent of such mating attempts is the male successful in taking down the female, but males are extremely persistent.

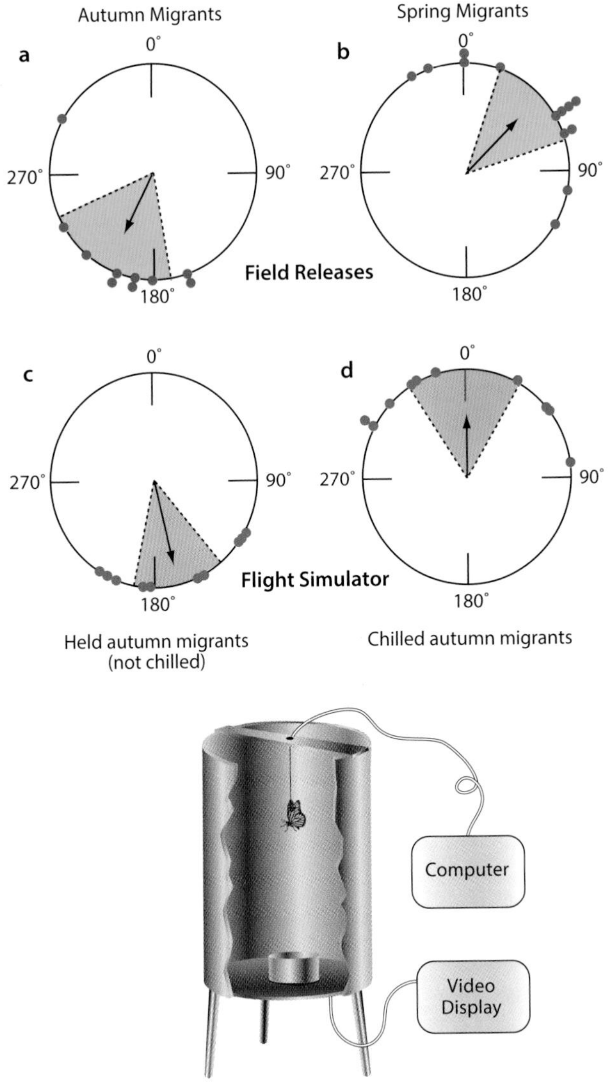

FIGURE 4.3. Compass orientations of flying monarchs under different conditions. (a) and (b): Field releases of adult monarchs and their flight paths at the time of disappearance from eyesight. Autumn migrants (caught in Minnesota) and spring migrants (caught in Texas) showed the expected patterns of southward and northward flight paths, respectively. (c) and (d): In more controlled trials with autumn migrants placed in a flight simulator, when monarchs were held under typical autumn-like warm conditions in the laboratory for four months, they continued a southward migration (c). Nonetheless, when held in the laboratory under more chilled conditions (like those experienced at the overwintering colonies), monarchs reversed their flight path, orienting northward (d). Within each compass, every dot is an individually tested butterfly, and the arrow indicates mean orientation of the group (shaded areas are 95 percent confidence intervals).

I should note that this section of text describing the mating behaviors of monarchs runs the risk of being a bit over-anthropomorphized. Biologists often struggle with how to interpret an animal's behavior without an excessively humanized viewpoint. Here is where both natural selection and comparisons between species can be very useful. First, where possible, our gold standard should be to interpret specific behaviors with respect to their particular fitness consequences (are the behaviors actually beneficial in terms of having successful offspring?). And second, if the behaviors of closely related species are distinct, is there a hypothesis for why natural selection may have resulted in divergent behaviors? Although our human lens is really the only one we have to study animal behavior, a framework of natural selection and evolutionary comparisons helps us to interpret the behaviors somewhat more objectively.

Back to our butterflies. After the pair settles on a perch (or on the ground), either with or without incident, courtship proceeds by the male orienting himself next to the female (fig. 4.4). Again, although this is sometimes accomplished without force, the male is typically clutching and holding down the female. Then, the abdomens (the last segment of the insects) connect, and copulation occurs, usually with the male opening and closing his wings repeatedly. This is followed by a postnuptial flight, where the male flies with the female passively attached, often to a somewhat secluded area such as dense vegetation. But mating has not yet been consummated. The postnuptial flight is critical because the butterflies stay together, usually for hours, and it is often after dusk when the sperm packet (called a spermatophore) is fully secreted into the female's pouch (termed the bursa copulatrix). In the case of the monarch, a spermatophore transferred from the male to the female can weigh as much as 10 percent of his body mass, and is largely a mass of nutrients to fuel the female and feed egg maturation. A small sac at the very back end of the spermatophore is the package of sperm that, if successful, will fertilize the female's eggs. It is a long process, with many steps, and with several chances to fall apart.

Both male and female monarchs mate multiple times. It is generally believed that, for males of almost any animal, more matings will enhance his genetic fitness. This is termed Bateman's principle, which is based on the notion that sperm are relatively cheap, and therefore spreading them around is often beneficial in terms of siring more offspring (fitness). For females, however, more matings are not necessarily better, since she is typically limited by

FIGURE 4.4. A typical sequence in monarch mating: (a) The take down from the air, followed by (b) a struggle on the ground, (c) *in copulo*, with the male dragging the female in flight, and then (d) back-to-back with the transfer of the sperm packet typically occurring after several hours. Not all monarch matings occur in this coercive fashion, but it is the most typical mating routine.

the number of eggs she can produce, and the sperm from a single male is usually plenty to fertilize all of her eggs. But, for butterfly species where males produce large spermatophores, as in monarchs, the situation is a bit more complex, because the "cheap sperm" part is only a small fraction of the spermatophore. It is in large part composed of a nutritional gift. Monarch males can mate upwards of ten times before their stores are exhausted. Females benefit from receiving multiple spermatophores, in part because of the high protein content of the gift. This protein can be food and fuel for female monarchs, both of which are critical at the end of their long overwintering period. Indeed, more or heavier spermatophores from males contribute to greater egg production in females. Although a monarch's reserves are dominated by energy stored from being a caterpillar eating milkweed leaves and from converting nectar sugars to lipids, spermatophores are a critical additional resource for females. Nonetheless, there are costs of multiple matings, the most serious of which is that with too many matings, an inserted spermatophore can cause the female's bursa copulatrix to burst, which results in death. Females need sperm to lay eggs, and spermatophores for nutrition, but too much can be deadly (fig. 4.5).

What I find surprising about monarch mating is not the apparent struggle or the different interests between males and females. After all, nature is full of crazy sex. The unexpected and interesting part is that monarchs have changed their ways, evolutionarily departing from the norm of most other milkweed butterflies (in the tribe Danaini). Among the 170 or so milkweed butterflies, two aspects of their mating behaviors are common to nearly all that have been studied except monarchs. First, most of these butterfly species have very complex mating pheromones (chemicals that affect the behavior of other individuals of the same species). Adult males typically first acquire toxic substances (not cardenolides, but pyrrolizidine alkaloids) from plants by scratching leaves and slurping up the contents. These plants are unrelated to milkweeds and are not used as food by the larvae. After substantial biochemical rearrangements, these alkaloids are then used by males in a pheromone to lure females during courtship. Second, it is common among many butterfly species, especially in the Danaini, that females are highly choosy about which males to mate with. Despite the many commonalities among the Danaini, monarchs stand out for not following the general trends. Indeed, monarch mating is not typical of milkweed butterflies, and there is little evidence for pheromonal communication or female choice.

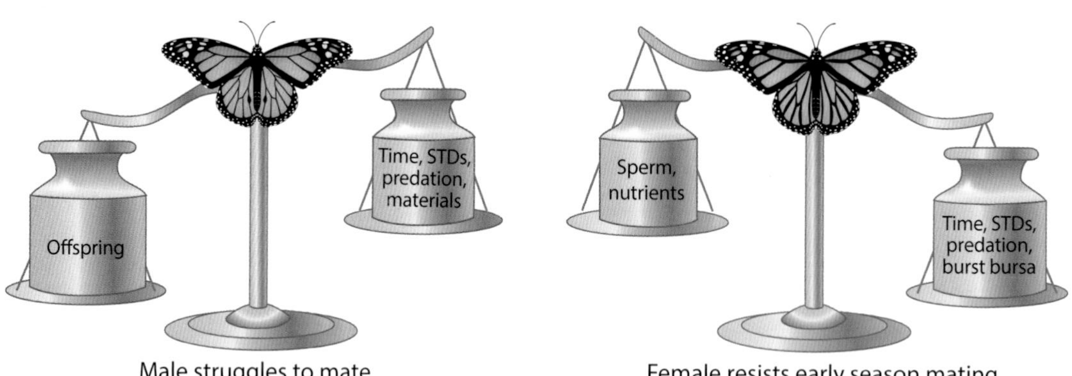

FIGURE 4.5. The evolutionary dilemma of multiple matings and the costs and benefits for monarchs. According to a model by Oberhauser and Frey, males attempt to mate, even if the female is not receptive, because the benefits of siring offspring outweigh the potential costs of sexually transmitted diseases, predation, and investment in the spermatophore. In contrast, females are potentially burdened by the time investment, potential damage, and other costs that favor resisting too many matings. Females often resist mating attempts by males, especially in the overwintering grounds before their northward spring migration.

Karen Oberhauser, a leading monarch biologist from the University of Minnesota, has hypothesized that the male monarchs' coercive means of mating evolved because of their unusual overwintering biology, which involves months of diapause followed by an additional lengthy migration back to the breeding grounds of the southern United States in spring. Mating begins in Mexico, but continues through the spring migration. Nonetheless, many monarchs that make it to Mexico simply do not have the energy reserves (stored fat) to undertake the return migration, some seven hundred miles, and hence it may be in the weakest and most worn males' interest to coercively mate before the migration commences. Even though early matings sire disproportionately few eggs (because early inserted sperm is displaced by later mating), it may be the best hope for weak males that cannot make the spring migration. For females, especially at the overwintering colonies, there may be some benefit to receiving spermatophores from early matings (in terms of fuel), and there may be relatively few costs to one or a couple of matings before taking flight for the spring migration. Nonetheless, there are high costs of too many matings for females (see fig. 4.5), who accordingly prefer to mate on their way north and once in the southern United States. Thus, the evolutionary departure of monarch mating compared with other milkweed butterflies may be linked to their overwintering biology. Chapter 8 will consider what factors drove their southern migration and overwintering biology in the first place, but for now, let's continue to follow their northward movements.

SINGLE SWEEP OR SUCCESSIVE STEPPING-STONES?

For over a century, scientists have speculated that monarchs had both an autumn and spring migration, but the hows and wheres were not known. As early as 1904, John Henry Comstock and Anna Botsford Comstock, the founders of Cornell University's Department of Entomology (the first university department entirely focused on the study of insects), wrote about what we now call the stepping-stone hypothesis: "The mother butterfly follows the spring northward as it advances as far as she finds milkweed sprouting; there she deposits her eggs, from which hatch individuals that carry on the journey, and in their turn lay their eggs as far north as possible. Thus generation after generation pushes on until late in the season we hear of them as far north as Hudson Bay." Yet it took another fifty years before major strides were made in understanding the spring migration.

The protagonists of this part of our story are Fred and Nora Urquhart, who should be designated as the academic foreparents, not only of monarch biology and migration, but also of the "citizen science" movement. Fred was a child-entomologist, and while a graduate student, in 1937, he made initial attempts to study monarch migration by marking individual butterflies as they flew south in the autumn. By 1940, he had developed a tagging system. He affixed a small label to the wings, with the hope of recapturing them later to understand their flight path and destination. Although interrupted by his service in World War II, tagging efforts continued in the early 1950s. After the war, Fred and Nora married and collaborated at the University of Toronto to decipher the biology of monarchs, with a special focus on where they went in the winter. They realized that to really understand the monarchs' movements, they would need to enlist the help of butterfly enthusiasts all over North America. Monarchs were found over millions of square miles, with known populations in Canada, all across the eastern United States, and in California. In 1952, Nora Urquhart wrote an article published in *Natural History* magazine (although only Fred was credited), outlining the knowns and unknowns of monarch ecology. And at the end of the article, she added the following:

> In order to obtain sufficient information to answer the mystery of the migration of the monarchs, thousands of specimens must be tagged in as many localities as possible during the northward migration between the months of April and June and on the southward migration between August and October. It is hoped that some of the readers of this article will want to assist in the tagging and in the rearing experiments. If you want to take part in this project, address your letter to: The Royal Ontario Museum of Zoology and Paleontology, 100 Queen's Park, Toronto 5, Ontario. Necessary labels and paper punch will be mailed to you at no cost. If enough people cooperate, we may someday be able to tell the complete story of this mysterious traveler.

The respondents initially numbered fewer than a dozen, but this grew to more than two hundred by 1955, and into the thousands by the early 1970s. As we will see later, this engagement of citizens in science was critical to the discovery of the monarchs' overwintering grounds and for fueling a culture of public interest in scientific knowledge. But this endeavor required more than twenty years of persistence following the publication of the 1952 article in *Natural History*. In the meantime, the Urquharts cultivated the growing group of citizen scientists, synthesizing their independently collected knowledge and data into formal reports that were sent to all participants. The first of these, published in July 1955, and mailed to citizen scientists in thirty-eight states and five Canadian provinces, included color images, maps, important bibliographic references, and a healthy dose of scientific speculation (fig. 4.6):

> This is a report written primarily for those who have so generously assisted our Museum in the studies of the Monarch Butterfly during the past three years. For this reason (since it is in the nature of a personal report rather than a scientific disclosure) I have taken the liberty of theorizing, in some instances, in the hope that such theories will stimulate your thoughts and your imagination. It may take many years of careful observation in order to arrive at "proven" answers to the questions: "Where do the Monarchs go in the winter-time?" and "Do they return to the land of their birth?" Then again, by good fortune and assiduity, we may arrive at the correct answer this coming year (1955–56). For the present, I take this opportunity to thank all those listed herein for their assistance, not only in tagging specimens, but in submitting valuable observations and most helpful suggestions.
> Sincerely, F. A. Urquhart

Although Urquhart calls this a "personal report" it was a highly scientific forty-page document, made accessible to the interested reader. By providing the mailing addresses of each of the participants, Urquhart had hoped to encourage dialogue among the citizen scientists. This report ultimately morphed into a complete synthesis of what they knew about monarchs and their migration in a 1960 book *The Monarch Butterfly*. In the report and book, Urquhart introduced various theories, mostly wrong, about how and where monarchs spent the winter. Yet, there is an important point about this moment in the history of the monarchs' story. It has to do with the general issue of "known unknowns" and how scientists cope with critical pieces of missing information. In some cases, scientists proceed, ignoring the unknown. In other cases, we must speculate on the missing link, as we cannot otherwise proceed with research. The Urquharts simply did not know where monarchs went in the winter until 1975. And yet they were trying to piece together parts of the life cycle, including the spring migration, which required speculation (hypotheses to be tested) on their overwintering.

In 1960, Fred Urquhart's best guess was that monarchs spent the winter in the Gulf states as free-flying adults (and not in tightly packed, huddled roosts)

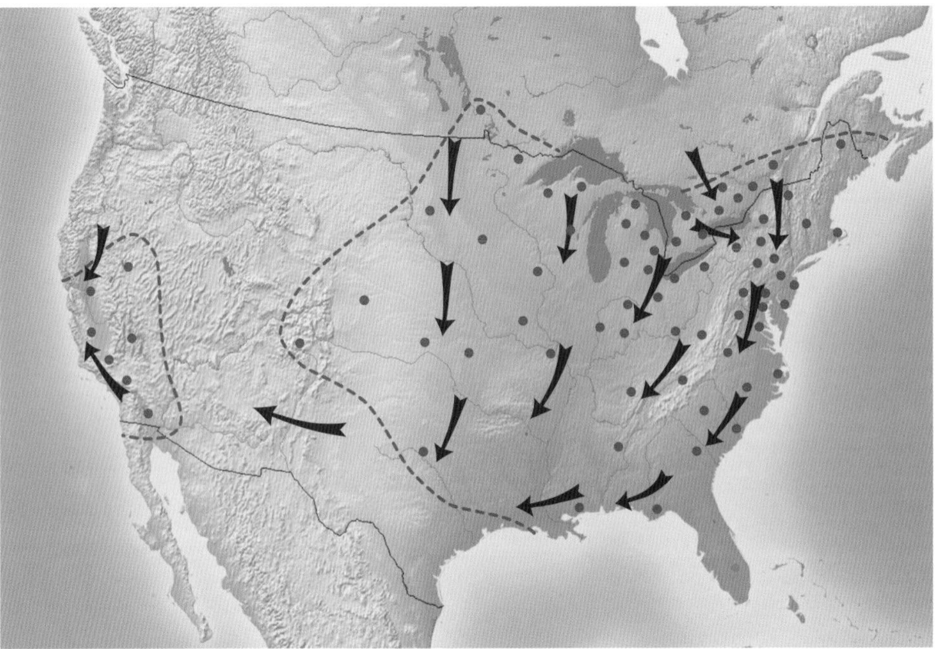

FIGURE 4.6. Sampling sites of citizen scientists between 1952 and 1955 (dots) as well as the hypothesized migratory pattern of monarchs by Fred Urquhart (redrawn from Urquhart's first report on the studies of the movements of the monarch butterfly).

and ultimately migrated to the Pacific coast. The Urquharts knew of roosting colonies of dense butterflies covering trees in California, and they suspected that free-flying butterflies from Florida and northern Mexico made it to California to finish overwintering before returning east in the spring. Despite tremendous tagging efforts, however, Urquhart's citizen scientists recaptured very few marked butterflies during their return in the spring. Urquhart outlined this hypothesis in his 1960 book, now dubbed the single sweep hypothesis. He was well-aware that it contradicted earlier "stepping-stone" ideas proposed by other entomologists, and in some ways he was a contrarian, happy to buck the dogma.

Urquhart's single sweep hypothesis suggested that autumn-migrating butterflies returned to the summer breeding grounds in the Midwest and Northeast (probably from California) in one fell swoop. Alternatively, the earlier introduced idea of the stepping-stone hypothesis advanced the notion that there were geographically intermediate generations of monarchs, and that butterflies migrated north until milkweed was available, and then laid eggs without further migration. Once mature, this next generation of monarchs would then push further north following the emergence of young milkweed. Given that monarch eggs hatch in a few days, caterpillars feed for ten to fourteen days, and metamorphosis from chrysalis to butterfly takes about a week, each month-long butterfly generation could be a step northward over the spring and summer. Yet when relying on observations alone, only limited progress could be made toward unraveling this mystery, especially because of the large geographical gaps where no citizen scientists were tracking and reporting monarch activity.

At this stage, the field had two leading hypotheses, but both were limited by the tagging method. The acute limitation of butterfly tagging is that massive people-power is needed on both ends of the migration. That is, even tagging hundreds of thousands of south-flying monarchs in autumn would result in far less than 1 percent of the overwintering population being tagged. Between the vagaries of butterfly mortality, different flight paths, and the limits of human observation, only a miniscule fraction could ever be recovered during the spring migration. In addition, spring migrating monarchs do not fly in large clusters, but rather individually, and the availability of milkweed throughout the southern United States leads to a rather diffuse process of remigration. Any process that requires marking and recapturing individuals at such a scale could not reliably reveal the details of such a complex annual cycle.

What was needed was a method by which a reliable signature of the origins of monarchs could be detected, without the need for such a monumental workforce on both ends of the migration. To understand the natal origins of a butterfly, we must be able to estimate where the caterpillar stage had been feeding. This was especially important in understanding the spring migration, as butterflies captured in the summer could either have come straight from

Mexico or from a generation born in the southern United States. Was Urquhart right, that it was a two-way migration made by the same individuals? Or were there discrete stepping-stones by which the monarchs migrate in the spring over several generations?

At this point in our story Urquhart and Lincoln Brower crossed paths. In the years following Lincoln Brower's contributions to the understanding of monarch sequestration of cardenolide toxins (detailed in chapter 2), his lab and widespread collaborators would continue to decipher the details of cardenolide chemistry moving up the food chain from plants to monarchs, ultimately impacting their bird predators. As we will see in chapter 8, Urquhart and Brower butted heads bitterly about the discovery of the overwintering grounds. Nonetheless, it was their use of complementary methods that finally began to yield answers to the problem of the spring migration. In particular, Brower's team was able to empirically advance the stepping-stone hypothesis by using milkweed's chemistry as a fingerprint.

In a classic research paper, Brower's team wrote: "From cardenolide analysis of individual butterflies in a migrating population, it may be possible to match individuals with their food sources from distinct geographic regions." In other words, if different milkweed species have restricted geographic ranges and produce different types of cardenolides, and if cardenolides were faithfully sequestered by monarchs, then an adult butterfly's cardenolide "fingerprint" would reveal where the caterpillar stage had been feeding. And fortunately, Robert Woodson, who was curator of the Missouri Botanical Garden's Herbarium, and a fanatical milkweed biologist, had developed range maps for nearly all of the more than one hundred *Asclepias* species in the 1950s. Consequently, it was possible to trace the plant-derived cardenolides up the food chain, revealing the natal origins of butterflies along the migration.

Over the next fifteen years, cardenolide fingerprints were used to cement the stepping-stone model. Brower's team (led by Stephen Malcolm, one of the gurus of milkweed chemistry and monarch migration) collected seasonal arrivals to the southern United States in April and early May, and more northern arrivals in the Midwest and Northeast through June. Based on the cardenolide fingerprint (as well as estimates of wing wear and modeling the monarchs' temperature-dependent generation time), the bulk of spring migrants from the Mexican overwintering grounds (at least 80 percent) had fed as caterpillars on common milkweed (*Asclepias syriaca*) in the Midwest and northeastern United States during the previous summer. The cardenolide signatures further matched a successive stepping-stone pattern whereby a large butterfly cohort completed a generation in the US South. Indeed, greater than 80 percent of the first generation of monarchs in the new year, the offspring of the spring migrants, fed on the spider milkweed (aka green antelope horn) *Asclepias viridis,* a species from the south central United States. One additional finding revealed by these studies is that migrating monarchs, those in the "advance team" moving north early in the summer, do lay eggs on milkweeds as they fly north. Thus, although the successive stepping-stone model was supported, each successive generation is not completely discrete.

Despite the brilliance of the cardenolide fingerprint method, the accuracy of natal origins is regional at best and is dependent on the limited range of particular milkweed species. Although nearly all overwintering monarchs in Mexico had fed as caterpillars on *A. syriaca*, this plant ranges from southern Canada to the Carolinas and west to the base of the Rockies. The cardenolide fingerprint was not that geographically specific. Scientists needed something more precise to understand the spring migration. The next technological advance to enter the picture was the use of stable isotopes in the mid-1990s. With isotopes, greater accuracy is possible. Given that adult butterflies experience no further growth, and their bodies are composed of materials garnered as caterpillars (namely, leaves), the cardenolides and elemental isotopes in adults hold the secrets to their larval origins. To understand isotopes, let me first provide a very quick primer on atoms, the smallest unit of a chemical element.

An atom consists of a nucleus, composed of protons and neutrons, which is surrounded by swirling electrons. Although the number of protons is always the same for any particular element, the number of neutrons can vary, and it is this differing number of neutrons that creates an isotope. The isotope number is determined by the total number of protons and neutrons in the nucleus. Take two fundamental elements, hydrogen and carbon, both essential for life on earth. The bulk of hydrogen on the planet, 99.98 percent to be precise, comes in the form of ^{1}H: one electron and one proton, but no neutron. Nonetheless, other stable forms of hydrogen exist. Deuterium (^{2}H) has one electron, one proton, and one neutron, and is sometimes called heavy hydrogen. Despite the fact that ^{2}H makes up less than 0.02 percent of hydrogen on the planet, its ratio with ^{1}H consistently varies across latitude. In other words, climatic patterns, especially air temperature, determine the ratio of ^{1}H to ^{2}H, and this varies consistently across north to south continental gradients. Similarly, carbon isotopes vary in plant tissues longitudinally (from east to west). With these two elements alone, a latitudinal and longitudinal grid point can be identified to determine the natal origin of a butterfly.

The key to the success of isotopes as a marker for monarch caterpillar origins was introduced in 1998 by Leonard Wassenaar (then working for Environment Canada, and now with the International Atomic Energy Agency), and a few studies have appeared since then that continue to improve our understanding of their migratory patterns. In particular, a 2013 study using stable isotopes confirmed that the year's first new generation of monarchs completed development in Texas and southern Oklahoma, but nonetheless, some first-generation butterflies developed as far north as Missouri and southern Illinois (fig. 4.7). In other words, most spring migrants leaving the Mexican highlands lay eggs (and die) in the southern United States, but a few butterflies travel remarkable distances on the return journey from the overwintering grounds—up to 1,500 miles (2,400 km)—and make it to the lower Midwest.

Despite these advances, the details of the spring migration have been subject to controversy, and some aspects of the monarchs' migratory journey remain a mystery. Because of the continental scale of the northern flight, and because it is a temporally moving target through the spring, initial tagging efforts, observations of the extent of wing wear, and mathematical modeling were not

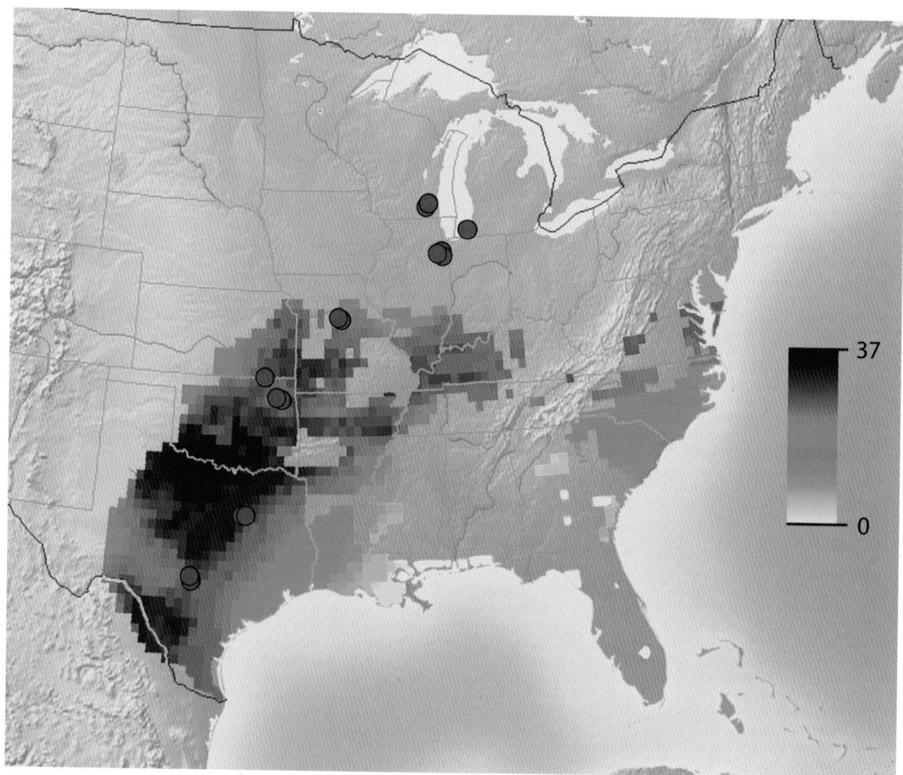

FIGURE 4.7. Origins of the new year's first generation of monarch butterflies as determined by isotopic ratios of hydrogen and carbon. Red circles indicate collection locations of adult butterflies in April and May (N = 78 butterflies). The white to green to blue gradient signifies the 2-to-1 probability of natal origin of those butterflies. For example, the darkest blue here represents that location having a probability of producing thirty-seven of the seventy-eight butterflies (nearly 50 percent) within the April–May collection. The bulk of the first generation of butterflies that later migrates to the Midwest and Northeast developed on milkweeds in Texas and Oklahoma.

conclusive. The successive stepping-stone model has received the most support, and it is clear that the bulk of spring migrating butterflies lay eggs and perish in the south-central United States. Nonetheless, some butterflies apparently go west and join the California population of monarchs; others may go as far east as Florida and the Carolinas and lay eggs (fig. 4.8); and still others make it as far north as the lower Midwest. Then, the bulk of the next generation

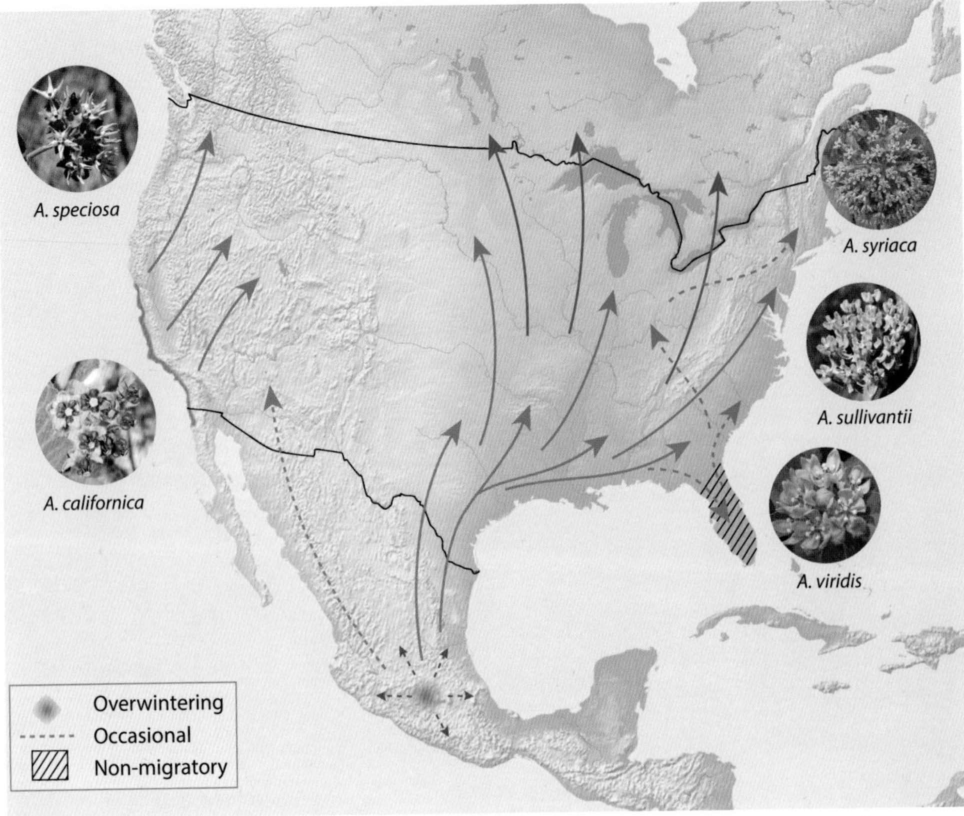

FIGURE 4.8. A schematic map of the spring migration of monarchs with images of major plant resources as part of this remigration. Note that the stepping-stone model is depicted by the migration from Mexico to the southern United States. The next generation moves further north. Caveats to this pattern are shown with dashed lines (some butterflies stay in Mexico, moving to the lowlands). Some butterflies also migrate to the southwestern and southeastern United States. The California population and migration is largely independent of that from the eastern population, although some mixing occurs. Resident monarchs also breed year-round in southern Florida and over much of Mexico.

moves further north. There are still many mysteries about the monarch migration to be solved. I suspect that climatic variability plays a huge role in determining when the monarchs leave Mexico and the extent to which they either play generational hopscotch or simply sweep across the continent as they move north.

FLUTTERBY PSYCHOLOGY

For monarchs, months of waiting, mating, and migrating have led to the beginning of the life cycle: the female's choice to lay an egg on a leaf. Monarchs are members of the butterfly family Nymphalidae, whose common name is the "brush-footed" or "four-footed" butterflies, because their front pair of legs is highly reduced; they look like little brushes, and these butterflies walk on only four of their six legs (fig. 4.9). Their middle and front tarsi (or feet) are covered in spines with sensory cells used to assess plant chemistry. Once a gravid female (one ready to lay eggs) arrives at a milkweed, her ritual assessment of a plant can be quite involved and often begins with the butterfly curling her abdomen, allowing her to touch the plant with her middle pair of legs, her brush-footed legs, and her antennae. Her tarsi (feet) may then rapidly tap or drum on the leaf surface, sometimes so fast that it is difficult to observe. She typically then drags her legs across the leaf's surface and taps with her antennae. She may also dab her ovipositor on the leaf, which also has sensilla and likely has photoreceptors to assess the color of the leaf. This entire ritual takes place before laying an egg. The sequence of exercises can be concluded within five seconds or, if interrupted, can take much longer. Interestingly, given the spiny nature of her legs, it is common for monarchs to scratch the plant's surface, potentially exposing the butterflies to chemical information hidden within leaves. Overall, this pre-egg-laying ritual involves chemical sensing of the plant using several different insect organs.

Lepidopterists have long wondered about when and why butterflies decide to lay eggs. It is perhaps the most important decision a female makes, and one that typically restricts her caterpillars to a single host plant for some time. When a former graduate student in my laboratory, Nile Kurashige, and I followed egg-laying monarchs on a hot day in southern Ontario, we decided we would become butterfly psychologists. As we joked about this in the field, we agreed that we were trying to get inside the head of a monarch to understand her egg-laying or "oviposition" decisions. And yet, like so many things in the

FIGURE 4.9. A recently emerged male monarch butterfly, with feet (tarsi) and brushlike sensilla showing on the legs. Note that the butterfly stands on four legs, with the front pair of reduced legs tucked up toward his eyes (but barely visible). The tufted black labial palps (part of the mouth) with two white spots are visible under the eyes and above the coiled proboscis.

natural world, the senses we use as humans may be less used by butterflies (and vice versa). Of course, we were looking with our eyes. As Tom Eisner and Jerry Meinwald, forerunners of the discipline of chemical ecology wrote in 1995, "Ours is a world of sights and sounds. We live by our eyes and ears and tend generally to be oblivious to the chemical happenings in our surrounds. Such happenings are ubiquitous. All organisms engender chemical signals, and all, in their respective ways, respond to the chemical emissions of others. The result is a vast communicative interplay, fundamental to the fabric of life."

What stood out to my student and me on that hot day was that monarchs in a stand of tall, robust milkweed plants seemed to be attracted to the taller plants and those with open flowers. Not surprisingly, they often stopped to take a drink. However, when fluttering between patches, they were preferentially laying eggs on single small milkweed plants. When not in dense patches, butterflies seemed to pass up healthy and mature plants, and instead were laying eggs on shrimpy plants—those with few leaves, because they had germinated only a year earlier or perhaps had resprouted after some catastrophic damage (like mowing) in the spring. As I have continued to watch over the years, it has become increasingly clear that female monarchs favor smaller plants, such as seedlings or late-season resprouts. Such smaller, tender plants would appear harder to find than big plants, and perhaps might not even provide enough foliage for a developing caterpillar. Nonetheless, research has confirmed these observations, showing that young plants, or those resprouting in the middle of the summer, are highly sought after by adult females. A recent study has also found that burning the Oklahoma prairie in summer results in milkweed resprouts that may be beneficial for monarch oviposition later in the summer.

In addition to preferring yearling and resprouted plants, monarchs will sometimes lay multiple eggs on these smaller plants. To understand how surprising this last observation is, I need to tell you the second thing that monarch butterflies seem to look for when deciding to lay an egg. Actually, it is what they seem to avoid: other monarch eggs and damage by monarch caterpillars. Especially in eastern North America, monarchs in natural field environments tend to avoid laying more than one egg on a plant. Published data and recent experiments conducted in my laboratory confirm that monarchs avoid plants with leaf damage and typically do not lay eggs on plants with more than one egg. This pattern has not been observed in all locations. For example, in Australia, where monarchs and their host plants are introduced organisms, plants often receive many eggs. The avoidance of occupied plants in northeastern North America is not surprising, however, as monarch larvae will cannibalize unhatched monarch eggs. In addition, higher densities of larvae on a plant result in low growth and survival, especially for young caterpillars. Despite these perils of multiple eggs, the look, smell, or other characteristics of young plants override any innate avoidance of competition, and monarchs frequently lay multiple eggs on young and resprouting milkweeds. It is yet unknown what specific cues from the plant might be mediating these preferences.

There are, however, specific chemicals, or "oviposition stimulants," that have been identified in milkweeds that promote egg-laying. These compounds, called quercetin glycosides, are in a class of chemicals called flavonoids, which are in all plants and function as pigments in flowers, as sunscreens to reduce ultraviolet damage in leaves, and as plant defenses against microbial diseases. Incidentally, many flavonoids are peddled as being healthy compounds for humans, but the Food and Drug Administration has cautioned against such healthful claims, at least as of yet. Two specific quercetin glycosides, in particular $C_{21}H_{20}O_{12}$ and $C_{26}H_{28}O_{16}$ are enough to stimulate monarch oviposition, even if applied to a wet sponge. Monarchs sense these compounds using sensilla on their middle and front pair of legs and antennae. Plants of preferred milkweed species tend to have higher amounts of these quercetin glycosides, and young leaves have more of these compounds than do older leaves.

So, why do milkweeds produce quercetin glycosides, specific compounds that stimulate the oviposition of its herbivorous enemy? You may recall from previous chapters that milkweeds don't need monarchs—not as pollinators, nor for any other known reason. So, although they live in very close association, evolutionary theory would predict that natural selection should favor milkweed plants that do not produce these oviposition stimulants. The likely scenario is something like the following. Quercetin glycosides are essential plant compounds, serving in some important role, likely as a sunscreen or in defense against microbial pathogens. Given this essential role, their continued production is advantageous. Even if natural selection by monarchs would favor reduced levels of quercetin glycosides, the plant is trapped between a rock and a hard place, perhaps the undesirable position of deciding between defense against damage by ultraviolet light or by monarchs. Such ecological tradeoffs are commonplace in nature. For example, as discussed in chapter 2, traits of flowers often are attractive to both pollinators and enemies of plants. In the coevolutionary battle between monarchs and milkweeds, trade-offs are critical as limitations on adaptive evolution. For egg-laying decisions, plant chemistry and the monarch's assessment of plant chemistry is central.

A TOXIC CHOICE

Here is where oviposition decisions come to intersect, not only with milkweed's chemistry, but specifically with cardenolides. Among all of the other factors that influence oviposition, when butterflies are within a cluster of a single milkweed species, with roughly similar attributes, egg-laying females choose plants with intermediate levels of cardenolides. In a series of beautiful studies led by Myron Zalucki of the University of Queensland in Australia, he and coauthors showed that female monarchs prefer to lay eggs on plants within a somewhat narrow window of cardenolide levels (between 2 and 4 mg of cardenolides per gram of dry leaf mass). These same intermediate levels allow high sequestration by caterpillars as well as high survival of the hatching caterpillars through the end of their first instar (which is the name of the caterpillar life stages, delimited by molting their exoskeletons). If oviposition occurs on plants with higher levels of cardenolides, caterpillars suffer toxicity and may perish. Alternatively, when laid on plants with low cardenolides, caterpillars grow just fine, but sequestration of cardenolides is low after feeding on these plants, exposing them to greater risk of predation (fig. 4.10).

In the context of "eat and avoid being eaten," monarchs delicately balance food quality (lower cardenolides) and sequestration (higher cardenolides). Even for the highly specialized monarch, which eats only milkweed, adult females make decisions that will strongly impact their offspring's ability to eat and fend off predators. And as noted in chapter 3, populations of milkweed harbor substantial genetic variation in how much cardenolides their leaves produce. So how do monarchs sense the plant cardenolide levels such that they choose intermediate levels? Although I speculate on this below, it is an unsolved mystery. Despite the fact that monarchs can sense quercetin glycosides (oviposition stimulants that reside on the leaf surface) using their feet, antennae, and ovipositor laden with sensory cells, cardenolides have not been reported on the leaf's surface. Like quercetin glycosides, they are not volatile, so they are not wafting from the leaves. And, finally, as far as we know, the amounts

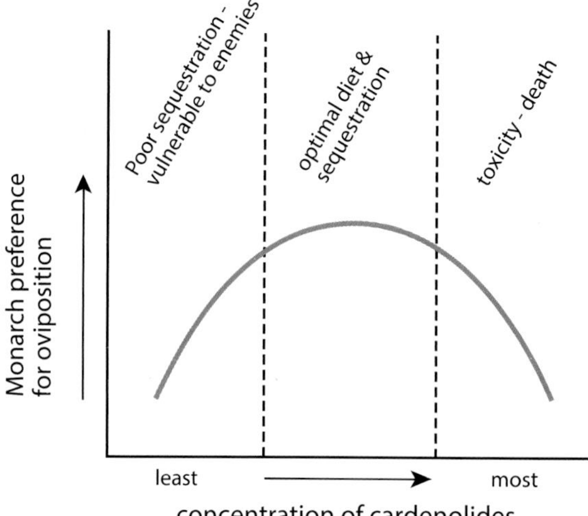

FIGURE 4.10. A conceptualization of adult female monarchs' preferences to lay eggs on milkweed plants with intermediate cardenolide levels.

of quercetin glycosides on leaves are not an indication of the amounts of cardenolides in leaves. Consequently, it remains a mystery how monarchs sense plant cardenolide levels.

There are two possibilities for how adult monarchs may sense cardenolides. First, if milkweed plants are in flower, their nectar contains small of amounts of cardenolides, concentrations that are reflective of the amounts found in leaves. Although the hows and whys of this relationship are unclear, here is what we know. Cardenolides occur in all plants parts, including the phloem, or plant plumbing that moves around the products of photosynthesis (primarily sugar). Not surprisingly, levels of cardenolides in different plant parts are often correlated. A plant with lots of cardenolides in its leaves will also often have high levels of cardenolides in roots and floral nectar. Thus, even as a passive process, when sugar is transported to floral nectaries, cardenolides may trail along. This is the simplest explanation for cardenolides in nectar. Nonetheless, cardenolides in nectar may provide egg-laying females with information about leaf cardenolides, providing honest information to flower visitors. A recent study in my laboratory by a postdoctoral researcher, Patricia Jones, confirmed that cardenolides in nectar can sometimes reduce bee pollinator visitation, but more strongly reduce oviposition by monarchs that take a drink before deciding to lay and egg or not. It is conceivable that milkweeds have evolved cardenolides in nectar to orchestrate reduced egg-laying by monarch butterflies.

Second, recall from the above description of the oviposition ritual that the monarch's tarsal spines often scratch the leaf surface. In other milkweed butterflies (in the tribe Danaini), such scratching is used to sense and imbibe plant compounds, but this has not been studied in monarchs. So, each of these chemical aspects of oviposition, including the quercetin glycosides, cardenolides in nectar, and leaf scratching, may hold further clues to the decision-making process of when to lay an egg. For now, suffice it to say that monarchs seem to be well aware of their host plant chemistry and this influences their critical decision of whether to lay an egg or not.

This chapter started with the waiting in Mexico, and then mating, and migrating northward in spring. And it has culminated in the behavioral ecology of egg-laying decisions. And yet, all this complexity leads us to the beginning of the monarch's life cycle. What happens when that egg, if laid, hatches? The subsequent biology that unfolds necessitates us to turn our thinking back to the coevolutionary arms race.

CHAPTER 5

Hatching and Defending

They were running hand in hand, and the Queen went so fast that it was all she could do to keep up with her: and still the Queen kept crying "Faster! Faster!" but Alice felt she could not go faster, though she had not breath left to say so. The most curious part of the thing was, that the trees and the other things round them never changed their places at all: however fast they went, they never seemed to pass anything. . . . "Well, in our country," said Alice, still panting a little, "you'd generally get to somewhere else—if you ran very fast for a long time, as we've been doing." "A slow sort of country!" said the Queen. "Now, here, you see, it takes all the running you can do, to keep in the same place. If you want to get somewhere else, you must run at least twice as fast as that!"

— Lewis Carroll, *Through the Looking-Glass*

After the monarch egg hatches on a milkweed leaf, it is all-out war. Most monarch caterpillars do not make it past the first day of life. And it is no wonder, given the diverse defensive barriers that the plant has erected. Just getting to the food can be difficult, and then when the caterpillar finally bites, it is like drinking from a fire hydrant. But it isn't water, or milk and honey that emerges; it is a toxic goo that will rapidly dry to a sticky mess. Despite these setbacks to the caterpillar, the plant has not won the coevolutionary arms race. Monarchs fight forward with offensive behaviors, their own physical armament, and enzymes that are fine-tuned to battle whatever the milkweed throws at them.

To imagine what these barriers are to a baby monarch caterpillar, visualize a toddler trying to eat a salad, with lettuce leaves covered in cactus spines and with a dressing made of thick and toxic glue. Scaling this up to a 150-pound adult human isn't pretty. Imagine yourself trying to eat soil, but having to first get down through a grassy lawn, one where the grass has not been mowed for weeks in the spring. Each blade of grass may be taller than your head, and you would simply snip each blade, one by one with your teeth, as close to the ground as possible. Once you have mowed, and are ready to sink your teeth into the soil below, you take that first bite, only to be met with a gallon of bitter glue welling up from what lies beneath. You will need to dig a mote with your mouth to drain away the bleeding bitter glue.

The quantification of such tit-for-tat matching of defensive and offensive tactics has led to important general principles in biology. Lewis Carroll's 1871 story of the Red Queen quoted above was used by the great evolutionary theorist Leigh Van Valen as a metaphor in a classic study, which he modestly titled "A New Evolutionary Law." In this scientific meditation, he derived the Red Queen's hypothesis, which was used to explain several important properties of the natural world. The one most relevant to the story of monarchs and milkweeds is that Van Valen theorized that most organisms are "running" to stay in the same place, that is, constantly evolving in response to competitors, predators, and parasites. Organisms pitted against each other stay, more or less, at pace with each other in their evolutionary arms race. Neither species appears to gain much of an upper hand.

This chapter focuses on how the results of this evolutionary arms race play out in the first days of life for a monarch caterpillar. Although adult females choose where to lay eggs, that is just the beginning of the ecological drama. During the short time following egg hatch, defenses and counterdefenses play out in a natural history theater presentation of insect behavior and plant reactions. We will explore the arsenal of traits and behaviors that each species has accrued as a result of its running to keep up. There are all kinds of plant defense strategies: kill the herbivores, slow them down, deny them nutrients, call in their enemies, or even just live with them. Milkweeds do it all. And monarchs cope with it all. And most of it can be observed early on in a monarch's life, beginning on the first day after hatching.

The first thing a hatched monarch typically does is to eat its own eggshell. Known as the chorion, the shell is largely made up of protein, and it provides a decent first meal for the monarch. As a young caterpillar, monarchs also engage in cannibalism. They eat other monarch eggs (but not caterpillars) that happen to be on the leaf. As explained in the previous chapter, an adult female will rarely lay more than one egg on a plant, and for good reason; the first one to hatch will typically cannibalize any neighboring unhatched eggs. But the cannibalistic phase is short-lived. Before monarchs begin their vegetarian ways, caterpillars first encounter two major plant defenses that prevent them from devouring a plant's leaves.

MOWING THE LAWN

Trichomes are leaf hairs. Although they annoy monarchs, trichomes may not have originally evolved for this defensive purpose (fig. 5.1). Plant species with a lot of trichomes are often found growing in full sun and hot, arid habitats. In such situations, trichomes function to shade the leaf surface from damaging ultraviolet light and reduce the heat load and water loss. As you might expect, in hot and arid environments, too much light and heat are problematic, and slightly fewer rays reaching the photosynthetic machinery is no problem. In mapping the development of trichomes in relation to the evolutionary history (or phylogeny) of milkweeds, work in my laboratory revealed that the abiotic environment (nonliving aspects of the environment such as sunlight, moisture, and nutrient availability) was the driver of trichome evolution. We hypothesized that the ancestor of all milkweeds (in the genus *Asclepias*) lived in hot, dry places, most likely in Africa, and was covered, not with trichomes, but with a thick layer of grayish wax (fig. 5.2). Waxes and trichomes are both leaf coverings and have similar physiological effects on photosynthesis.

FIGURE 5.1. Monarch caterpillars on three different *Asclepias* species with varying densities of trichomes. (a) A newly hatched monarch on *A. californica*, which is among the wooliest of the milkweeds. (b) On another California milkweed, *A. eriocarpa*, this third instar (stage) caterpillar shaves the dense hairs before feeding on the leaf. (c) Even on *A. syriaca*, the eastern common milkweed, trichomes are often dense enough (between 100 and 300 trichomes per square centimeter) that caterpillars, like this newly hatched individual, "graze the lawn" before sinking their mandibles into the leaf.

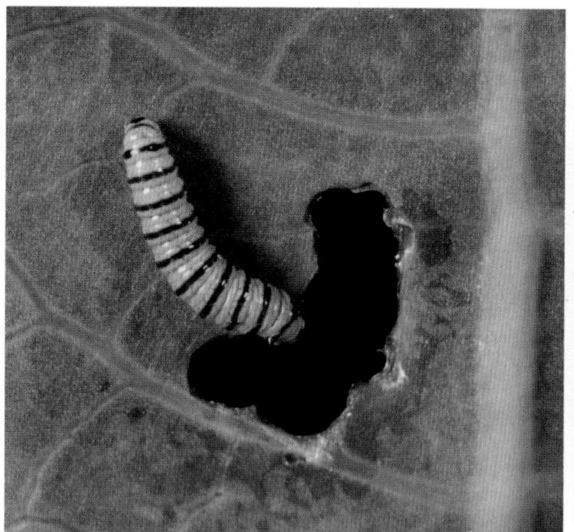
a

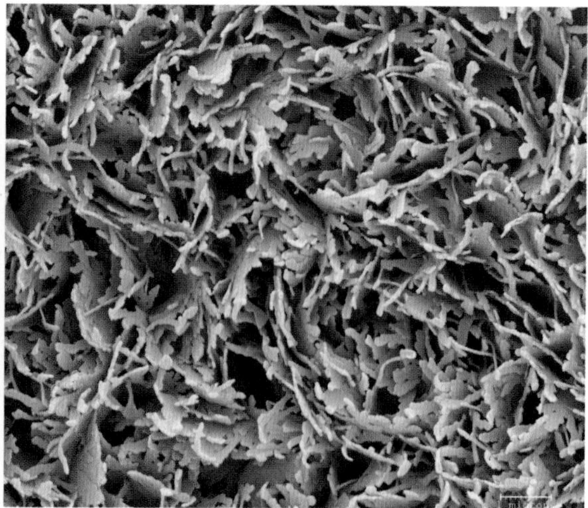
b

FIGURE 5.2. (a) Sandhill milkweed, *A. humistrata*, common in the southeastern United States, but which also grows in open, hot, and arid habitats, is completely free of trichomes. However, its leaves have a grayish appearance because of (b) a thick layer of wax crystals that cover the surface (scanning electron micrograph, zoomed in). These waxes function physiologically in a manner similar to milkweed trichomes, but they also make the leaf surface very slippery (monarch larvae can easily fall off). Plant species with trichomes lack the wax layer, and vice versa.

Although the wax layer physically protects leaves, as do trichomes, it has evolved discretely, like an on/off switch. Plants either have a thick layer or no wax at all. An evolutionary benefit of trichomes over wax is that natural selection can easily fine-tune the density and type of these leaf hairs to suit the environment. Later evolving milkweeds have lost the wax covering and have instead developed trichomes, perhaps owing to this advantage. And although their evolutionary origin is as a sunscreen, for a newly hatched monarch caterpillar, trichomes can be quite a challenge. As a counterdefense, monarchs shave those trichomes, as if they were grazing on a lawn. Young caterpillars cut the trichomes, but don't eat them, before they feed on the leaf. Although monarchs rarely die from the stress of shaving trichomes, this first barrier to feeding slows them down and exposes them to heat and predators. The next surprise milkweed has in store for young monarchs, however, can certainly kill them.

SO MANY LIVES LOST IN LATEX

In 1995, at the end of my first year in graduate school, I got some sage advice from my lab mate (and later my wife), Jennifer Thaler. "Learn to eat the study organism"—she wrote in a note to me. Fortunately, the plants and bugs I was studying at that time would not have done much damage. I have tried to heed this ecologist's mantra when possible. But it simply would not be safe or advisable for those studying milkweeds and monarchs. In chapter 7, I will discuss eating milkweed plants, several parts of which can be highly nutritious and delicious. But care must be taken in their preparation. And there is one rule for your culinary explorations of milkweed: never drink the latex.

On several occasions, I have inadvertently exposed my eyes and lips to latex. One of the hazards of being a milkweed biologist is the frequent contact with latex. Whether I am investigating potential field sites, harvesting leaves from experiments in the greenhouse, or simply taking a walk in the woods with my family, it is hard for me to pass up breaking a milkweed leaf to experience the latex. The latex of common milkweed (*A. syriaca*)—and likely all *Asclepias* species—is an extreme eye irritant. Although it is unclear what aspects of the chemical cocktail provoke the burning and tearing, it is rapid and unpleasant, and happens to me at least once a year. Perhaps of even greater concern was the time when I inadvertently touched my lips after bleeding the pine leaf milkweed (*A. linaria*) in my greenhouse at Cornell. In a matter of minutes, my lips were numb, and although they did regain feeling in the coming hours, I don't wish to repeat that experience. Needless to say, I have never encountered so much latex as to worry about being mired in it, or having the drying rubber glue my mouth closed. Yet, for a tiny insect species feeding on milkweed, such an experience is commonplace.

Latex is stored under pressure, and when a caterpillar finally takes a bite, puncturing milkweed cells, latex immediately spurts and oozes from the newly created wound. The recently hatched, hungry caterpillar faces the physical problem of sticky and quickly drying latex, and a new multifaceted chemical problem of the cocktail of poisons in that same glue. Probably the biggest threat to survival of a monarch caterpillar on its first day of life is latex. In a stunning series of studies, Myron Zalucki and his colleagues followed the fate of more than three thousand monarch eggs on nine different milkweed species in the field. Overall, more than 60 percent of monarchs died in the burst of latex that accompanied their first bites into the plant (see also fig. 1.2). Now that is a potent plant defense.

Because sticky white latex is such an important part of the biology of the "milk" weed, a general discussion of latex in plants is in order. As early as the seventeenth century, the term "latex" was used by English-speaking physicians (many of whom, as we have already learned, were excellent botanists). Its function was analogized to blood of animals, and it was studied in several distinct plant families. Not unlike the blood of animals, it travels in vessels; when the epidermal skin of latex-bearing plants is damaged, latex oozes out; and when exposed to air, it becomes sticky and quickly coagulates. The American botanist, Joseph James noted the repulsive and defensive qualities of milkweed's latex in 1887: "Serving as a vehicle for the conveyance of nourishment from the roots to the leaves, it carries with it at the same time such disagreeable properties that it becomes a better protection to the plant from enemies than all the thorns, prickles, or hairs that could be provided." Note that the first clause of James's sentence implicates latex as having a role in plant nutrition, an idea that has subsequently been debunked; nonetheless, he got the other half of the story right. By 1905, Hans Kniep, a German doctoral student showed experimentally that draining a leaf of latex made it palatable to otherwise deterred slugs. This landmark study was well ahead of its time because of the care and precision with which the experiments were conducted. As a result, the author demonstrated the only known function of latex more than one hundred years ago. Such experiments have been replicated numerous times with milkweeds in the modern era (fig. 5.3).

Not only is latex a visible and highly effective plant defense, but it is also one of the most common and widespread defensive traits in the plant kingdom. It

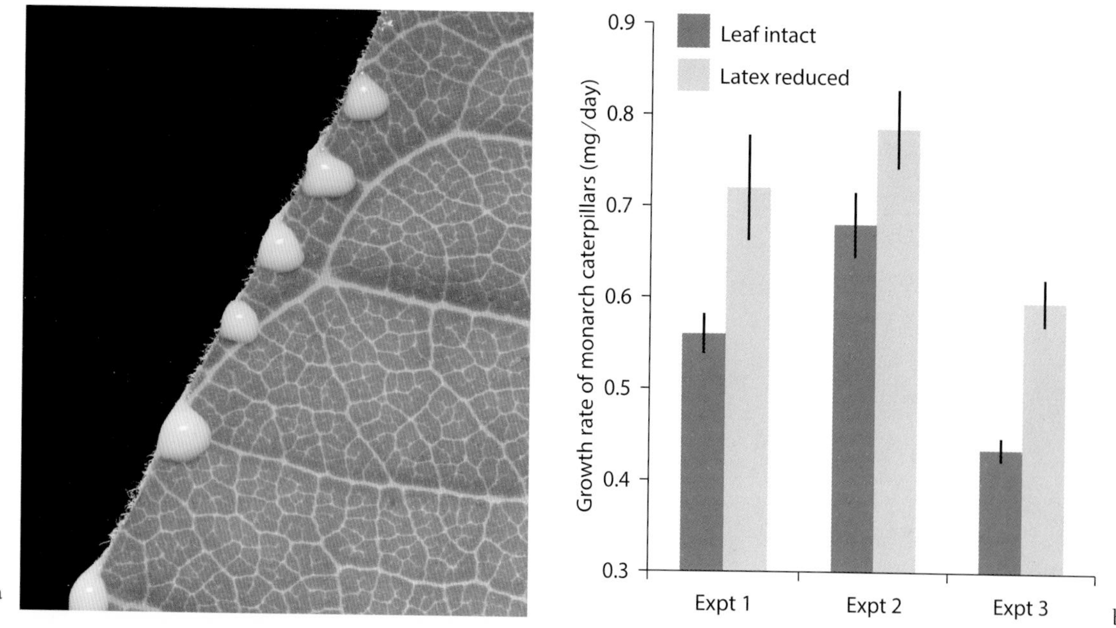

FIGURE 5.3. (a) Latex exuding from a cut leaf of common milkweed *A. syriaca*. Once a monarch caterpillar becomes mired in latex, death is common. Even when caterpillars survive, however, latex reduces their growth rate. (b) In three separate experiments with *A. syriaca*, when Zalucki and Malcolm experimentally reduced latex flow, the growth rate of monarch caterpillars increased substantially.

has repeatedly evolved, perhaps as many as thirty to forty times. Among the few hundred botanical families (Apocynaceae is but one of these), some 10 percent have independently evolved latex, and overall just under 10 percent of all plant species (likely more than thirty thousand species!) produce latex. Recall that convergent evolution is one of the hallmarks of adaptation, with the same traits repeatedly evolving toward the same function in diverse organisms. Even some mushrooms, in the genus *Lactarius* (aka the "milk caps") produce latex. This is a very common trait indeed (common to dandelions, fig trees, poppies, and even lettuce). And although explanations for the production of latex have included functions involving primary metabolism, namely, storage and movement of plant nutrients, waste, and maintenance of water balance,

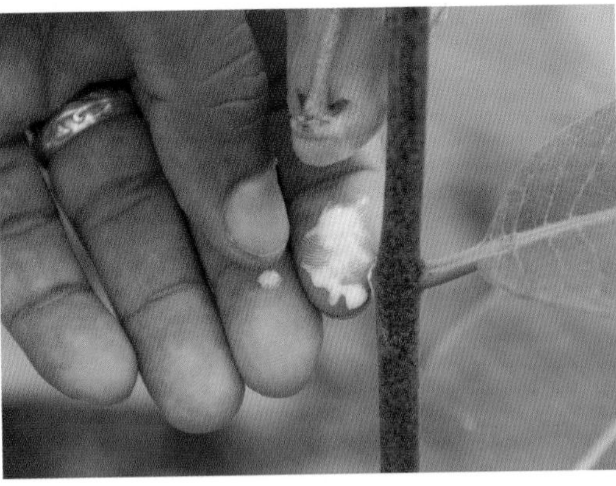
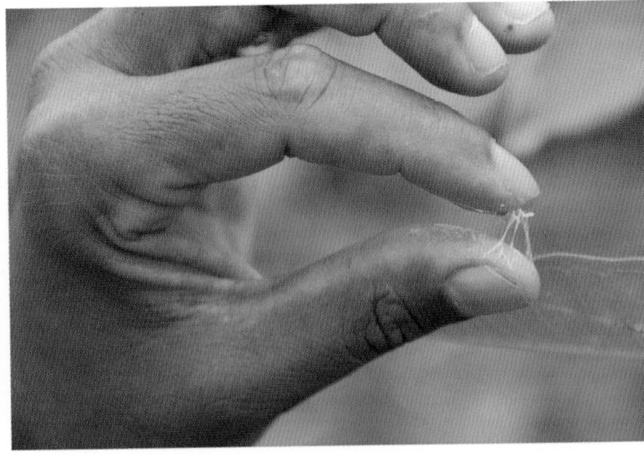

FIGURE 5.4. The author demonstrating that milkweed latex quickly coagulates into a sticky, rubbery mess—within a minute so.

none of these functions remain credible or have any scientific support. In other words, latex is first and foremost a plant defense.

Even though latex and its delivery system have repeatedly evolved, their chemical and morphological basis is quite similar among plants. Chemically, latex is a suspension or emulsion of particles in a watery fluid, usually under pressure in plant cells that form canals known as lacticifers. In the milkweeds, the canals follow the vascular plumbing of all aboveground parts of the plant (from stems to leaves to flowers and fruits), but they are composed of a mere sixteen cells, which elongate, branch, and spread to form the latex-delivering tubes throughout an entire plant. In other plant groups, but not in *Asclepias*, latex can also flow out from the roots. The compound that dominates the makeup of latex is rubber, derived from isoprene (a very common plant compound), and it can account for more than 40 percent of the composition of latex, depending on the species. Rubber is the key to the sticky nature of latex and gives it both the property of gumming up insect mouths, as well as sealing off plant wounds quickly. This sealing stops further drainage of latex and may prevent infection by microbial pathogens.

Latex is famous for its sticky properties (fig. 5.4), and it has been used to produce a wide range of products: rubber (primarily from *Hevea brasiliensis*

Euphorbiaceae); chicle from the genus *Manilkara*, which is used in chewing gum; and lacquers from phenols in the latex of plants in the mango family, Anacardiaceae. In addition to being sticky, various plant species' latex contains highly active secondary compounds, including the alkaloids used to make heroin from poppies. Latex also typically contains digestive cysteine proteases. Similar proteases are produced by animals to digest proteins; in latex, the apparent function of cysteine proteases is to eat away at the caterpillars' gut lining. The latex of milkweeds is no different; it contains cardenolides (up to 30 percent of the dry mass of latex), as well as defensive saponins, cysteine proteases, and several other compounds.

DISARMING THE LATEX BOMB

How does the poor caterpillar deal with the defenses the milkweed has thrown at it? Although more than 60 percent die during their first bites, many do make it through. So how do they do it? Under the rules of coevolution, each defense typically has an offensive counterpart. If a young monarch can successfully negotiate the trichomes and take a first bite, then it must deal with the latex outflow. Monarchs possess astonishing behavioral adaptations to disarm the latex munitions. I have termed the first adaptation the "circle trench" (fig. 5.5a), a moat around the monarch's castle. In a circle trench, the caterpillar has cut off the flow of latex to create an island upon which it will then feed, leaving a signature style of damage on the leaf (fig. 5.5b). Although monarchs do not initiate this type of feeding on all milkweed species, it is typical on the common milkweed (*A. syriaca*) and begins with a bite and latex outflow. Then, the tiny caterpillar often retreats, wipes on the leaf any latex that has come into contact with its head or legs, and it moves forward to continue digging the circle trench. It is interesting to note that monarchs sometimes dig the circle trench while shaving trichomes bit by bit as they go around (as in fig 5.5a), while at other times they shave much of the circle first and then begin the trench (as in fig. 5.1c above); both images show monarchs on common milkweed, *A. syriaca*.

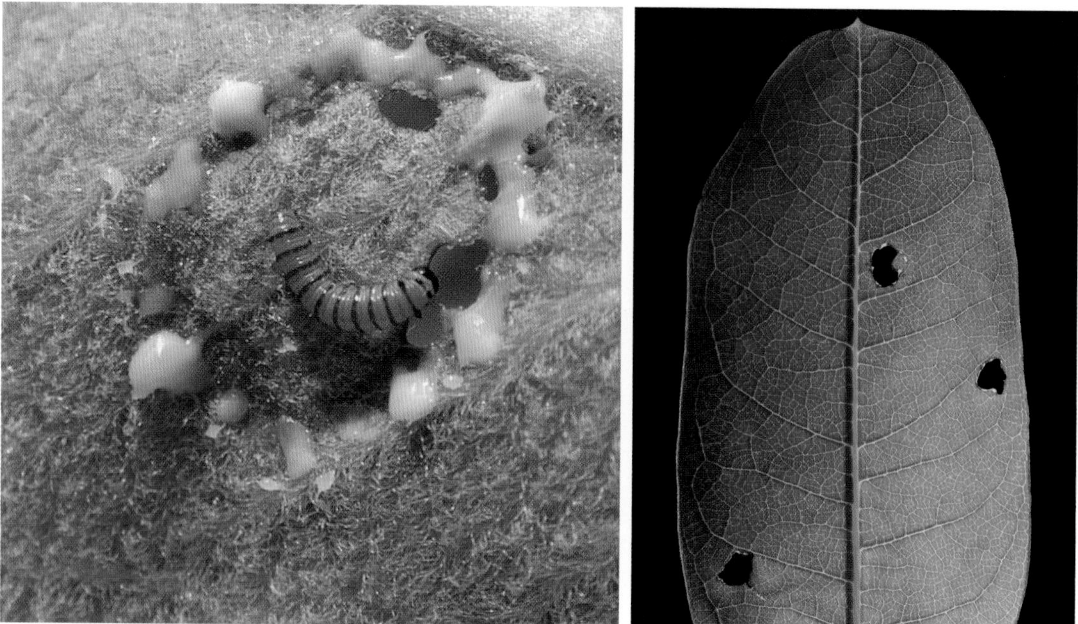

FIGURE 5.5. The first stage of monarch growth after hatching on common milkweed is chewing away leaf trichomes and digging a circle trench into the leaf. (a) This caterpillar, only hours old, is shaving trichomes in a small ring (less than a centimeter in diameter) and making bites into the leaf as it proceeds around that circle. Droplets of latex well up, and the caterpillar retracts its head before moving on and continuing to dig the circle trench. After the circle is complete, an island of milkweed leaf was ready for consumption without any latex flow. (b) In midsummer, in a field with many monarchs, one can often find leaves with these chewed-out islands. This type of damage is attributable almost exclusively to newly hatched monarchs.

In the first hours after monarchs hatch, this disarming of the plant's latex defense permits less than half of the caterpillars to survive. For those caterpillars that are less fortunate, catalepsis (a medical term used to describe being in a nonresponsive state) is a frequent outcome. As Myron Zalucki has said of monarchs on milkweed, "It's the first bite that counts"—and all too often, when substantial latex is imbibed upon the forceful ejection from a punctured leaf, monarchs enter a cataleptic state of motionless flacidity. If they are not overwhelmed and knocked out by the latex, monarchs can recover, usually in five to ten minutes, but it depends on their level of exposure.

Reducing the flow of latex is perhaps the strongest benefit for tiny, recently hatched monarchs to dig a circle trench. But there are other potential benefits too, and although these have not been directly studied for monarchs, two related milkweed butterflies studied in Borneo (*Leuconoe nigriana* and *Euploea crameri*) seem to gain protection from predation through their circle trenching behavior. These caterpillars bite the leaf, imbibe latex, and orally expel a foamy regurgitate around the circle trench. This barrier has been shown to fend off predatory ants. Monarchs do not produce such foam barriers, but perhaps the exuded globs of latex around the circle trench (see fig 5.5a) serve an equally protective role. Observations of predation attempts on monarchs (by ants or other enemies) inside and outside of circle trenches would be a fruitful avenue of research.

Many of the monarch behaviors discussed in this book so far fall into a category of what biologist call "phenotypic plasticity." An organism's phenotype simply refers to a trait or set of traits, including its physical attributes, chemistry, and behaviors. And plasticity is, well, just what it sounds like, flexibility. Formally, phenotypic plasticity is the ability of a single organism to exhibit different traits depending on the environmental circumstances. In the case of a newly hatched monarch, if a caterpillar is on a leaf with only a modest smattering of trichomes, it will not attempt to shave the leaf. Depending on other circumstances, monarchs may or may not dig a circle trench. As with the shaving of trichomes, the specific details of what will make a monarch trench a milkweed leaf have not been well-studied.

Take, for example, the few milkweed species that produce no detectable latex. Butterfly weed, *A. tuberosa*, which is native throughout much of North America, produces little if any latex, and young monarchs proceed with feeding on the leaves with no trench. Yet the Sandhill milkweed, *A. humistrata*, common in Florida, produces abundant latex, and this species presents different challenges as well. Although *A. humistrata* has no trichomes, its leaves have a highly waxy surface (see fig. 5.2 above), which renders the leaf slippery and makes young caterpillars vulnerable to falling off. Consequently, on *A. humistrata*, monarchs lay down a thick mat of silk before attempting to feed. This silk mat becomes a platform on which to rest and from which to deactivate the latex-delivering canals. Myron Zalucki and Lincoln Brower described the caterpillars' behavior on *A. humistrata* as follows: "Before feeding, first instars would lay down a mat of silk with characteristic side-to-side movements of the head (we saw all larval stages doing this). The larvae would then position themselves on the mat, usually with their heads towards the main midrib vein. With legs apparently anchored on the silk mat, the larvae either attempted to bite into the leaf, or struck the leaf surface with a rapid sideways and downward motion of the head. This mandible slashing caused an immediate outflow of latex, which formed into a small globule" (fig. 5.6). In turn, when successful, monarchs on *A. humistrata* do not end up making an island of safety from the latex, but instead they make arcs from their silk mat. Overall, their behaviors on *A. humistrata* are similar, but different, from those on *A. syriaca* (see fig. 5.5). Their stereotyped behaviors on the different plant species can be generated by taking an egg from *A. syriaca* grown in New York and placing it on *A. humistrata* grown in Florida (and vice versa). In other words, a single caterpillar is capable of changing its tactics depending on the milkweed it encounters, irrespective of where it was born. Indeed, their behaviors are phenotypically plastic.

I presume that the silk mat is much more important as a fixture point on slippery *A. humistrata*, whereas the mat is certainly not as important (or as apparent) on *A. syriaca*. The three behaviors—shaving of the trichomes, laying down a silk mat, and digging a circle trench—are all flexible, or phenotypically plastic, in newly hatched monarchs. To be clear, the differences in behavior are not due to genetic differences between individuals, but rather to the environment they experience. Because this butterfly species inhabits diverse regions of the Americas, each with distinct milkweed species, and it encounters many of these species at different times (and in different generations) of the monarch's annual migratory cycle, it has evolved the flexibility of eating any *Asclepias* species, despite the diversity of defensive traits that might be pres-

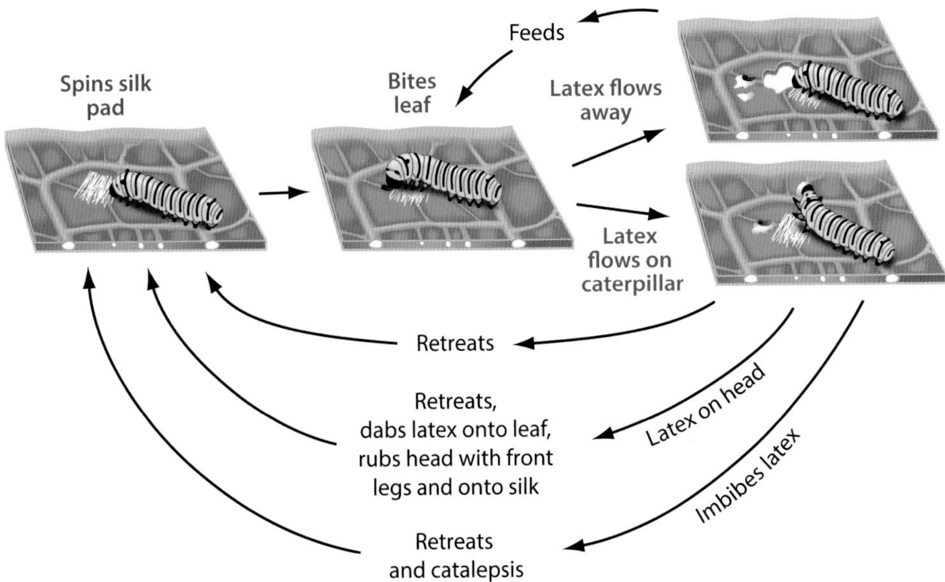

FIGURE 5.6. The sequence of behaviors associated with feeding by newly hatched monarch caterpillars on the leaves of the Sandhill milkweed, *A. humistrata*, common in Florida.

ent. Such plasticity, although widespread in nature, even in plants that I describe later in this chapter, is a tremendous evolutionary feat. Evolution by natural selection has gifted the monarch with the ability to, on the fly, match its tactics to the specific challenges it faces.

As monarchs grow, they continue to encounter and disarm latex, but their tactics change. More mature caterpillars, typically beginning in their fourth larval stage, cut off the flow to larger portions of leaves. Probably for many reasons, including the fact that the caterpillars are much larger and they need to consume exponentially more foliage as they grow, monarchs now either cut a notch at the base of a leaf's petiole (fig. 5.7a) or in the middle of its midrib (fig. 5.7c). In either case, like digging a circle trench, this may take tens of minutes, and often involves cutting, retreating, and wiping away of the latex

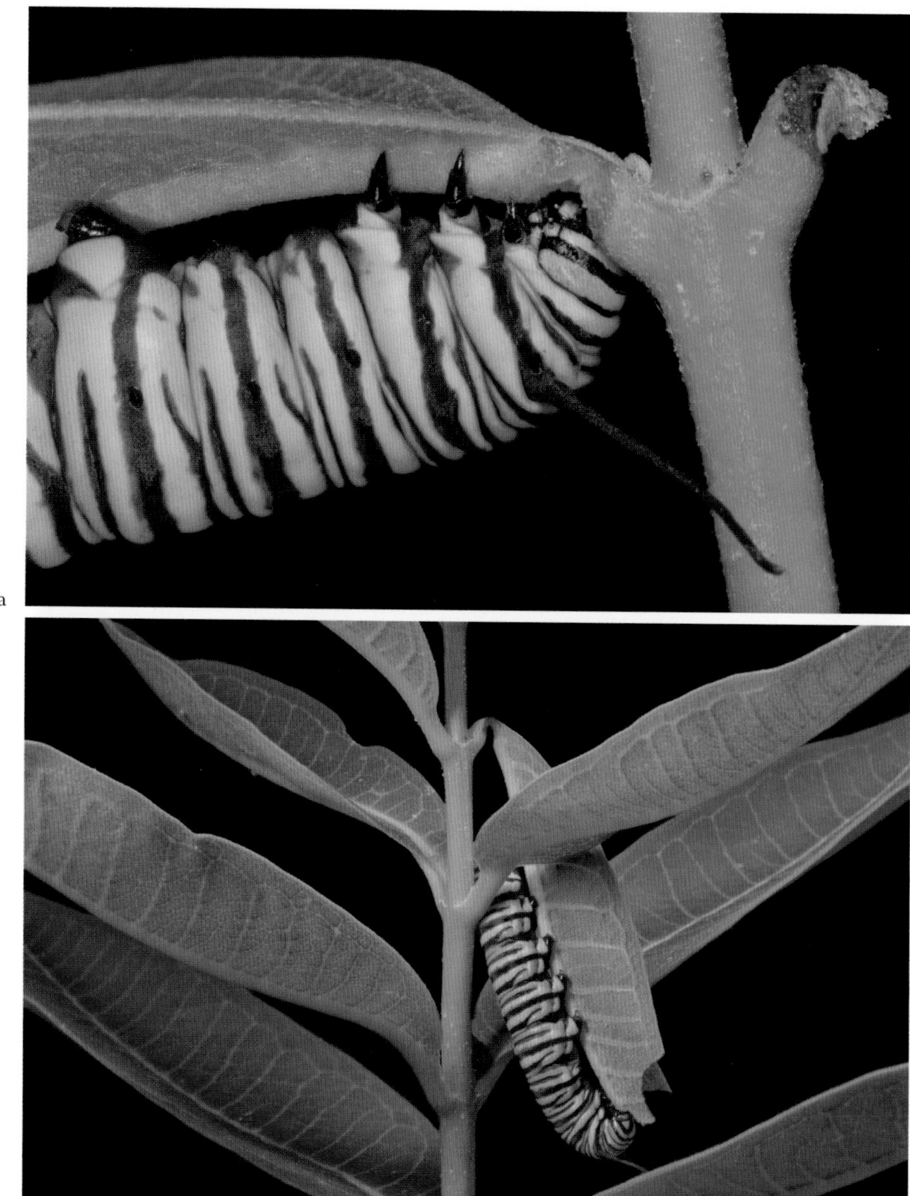

FIGURE 5.7. More mature monarchs also deactivate latex before eating leaves, but they cut off the goo at the source for each leaf. (a) and (b): A notch is being cut, after which the leaf hangs down, and the caterpillar flips around to eat the leaf. (c) and (d): Occasionally a large droplet of

c

d

latex wells up as the caterpillar is notching the leaf. In this case, the caterpillar may abandon the leaf, wipe its head free of any latex (shown), or, in rare cases, it may imbibe the latex and continue cutting the notch (not shown).

(fig. 5.7d). Ultimately the leaf hangs down, the caterpillar flips around, and usually it begins feeding voraciously on the disarmed part of the leaf right away (fig. 5.7b). Once again, the caterpillar feeds in the absence of the flowing latex.

The first published hypothesis about this behavior appeared in 1977 in *News of the Lepidopterists' Society*, where the editor, Jo Brewer, published pictures of how monarchs reduce latex flow and described the process: "Before beginning to eat a fresh leaf, these caterpillars chew through the stem of the leaf just far enough to cause it to droop, but not far enough to cause it to wither or to drop from the plant. . . . Obviously this cutting and drooping operation is in some way advantageous to the caterpillars. If it improves the nutritional value of the ingested leaves, the caterpillars which engage in it deserve an A+ in organic chemistry and hydraulic engineering. If it just happens that the caterpillars don't like too much gravy with the main course, they still deserve some credit for learning to turn off the faucet."

LIKE A CAT DRINKING MILK?

There are some monarch behaviors that I have witnessed only once. One behavior is rare enough that only a few people have observed it, and they have had quite different interpretations of the event. As unlikely as it may seem, the observations I made were over lunch in downtown Toronto, probably toward the end of the summer of 2002. At the time, I worked in a building called the Earth Science Center, a modern and very functional set of research laboratories in botany, geology, and forestry at the University of Toronto. Although sandwiched between the museum district and the "main" Chinatown in Toronto (there are three million people and six Chinatowns in Toronto), the university architects had the good sense to leave a little green space between the buildings and bustle. There lay a few hundred square meters, planted with trees, with a few benches, and some horticultural beds with ornamentals. This was a favorite place to sit and eat lunch when the weather was nice.

In addition to the plantings, various uncultivated plants grew abundantly in the garden, including some very large common milkweed. On this particular day, I ate my lunch by myself, and while sitting there I noticed a tall milkweed, more than six feet (two meters) at least, complete with a nearly mature monarch caterpillar on the underside of a leaf, probably five feet off the ground, in perfect view from my bench. What followed was no everyday sighting. As expected, the caterpillar began to bite the petiole of a leaf close to the stem. This is what often happens, and as described above, the various stages of monarch caterpillars have different approaches to deactivating the latex. Such a large caterpillar often makes a notch on the petiole until the leaf hangs down (see fig. 5.7). But, in this case, a large droplet of latex welled up in the notch as the caterpillar was cutting the vein. What the monarch did then was very odd: it maintained its head in the position at the petiole notch, which was filling up with latex. And as I watched, that droplet disappeared. I shook my head, in disbelief, but as I continued to watch, the caterpillar began again gnawing away at the notch and another large droplet appeared, and then, it too proceeded to disappear.

Watching latex exude is a real pleasure. I think it is because so few other plant defenses, save spines, thorns, and prickles, are visible, especially in the act of defending the plant. But, the disappearing latex was confusing. Why did the caterpillar not wipe the latex away? Why was it "drinking" the latex? What came to my mind at the time, and what I still believe, is that this was a necessary evil for that caterpillar. Monarchs are negatively affected by latex, but the only way for this individual to successfully deactivate the pressurized latex was to suck it up, in order to move the latex out of the way, especially while it was still liquid, and then continue to disarm the leaf before feeding on it.

Miriam Rothschild also observed this drinking of latex. In fact, in an obscure note she published (also in the *News of the Lepidopterists' Society*), Rothschild wrote that the monarch drinks latex like a cat drinks milk. Working with a colony of monarchs in England that was originally collected in Trinidad, Rothschild speculated that the caterpillars sought out latex because they needed to

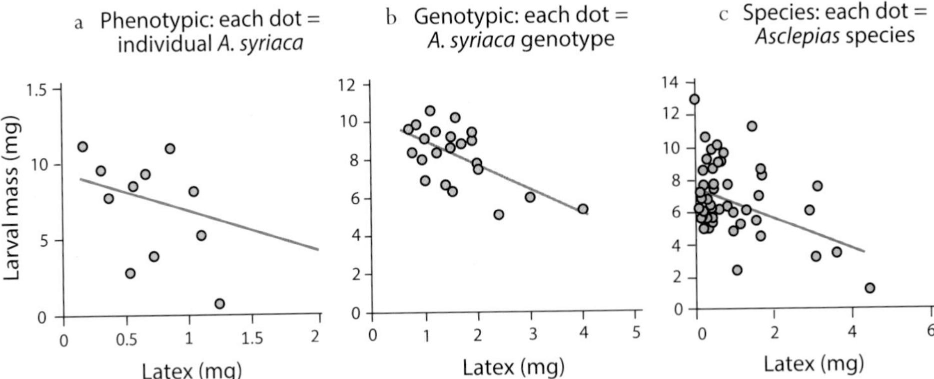

FIGURE 5.8. The most potent defense of milkweed is its milky latex. In all of our studies over the past decade, latex is the best and most consistent predictor of plant resistance to monarchs. These three different types of data all indicate the same pattern of reduced monarch caterpillar growth on plants with higher latex. (a) Each dot is a different individual of common milkweed, *A. syriaca*, grown in a growth chamber. The greater the amount of latex exuded, the slower monarch caterpillars grow. (b) Each dot represents a different genotype of *A. syriaca*, with several replicates for each genotype grown in the same field (means are shown for caterpillar growth and latex on each plant genotype). (c) Caterpillar growth on different milkweed species; each dot represents a different *Asclepias* species, again with means shown for growth and latex. The story is the same in each analysis: more latex means reduced monarch growth. Although each of these analyses is correlative, their consistency suggests that latex is indeed a highly effective defense against monarchs.

concentrate the cardenolide toxins within. Her observations have never been confirmed, and I tend to disagree with her interpretation. I would suggest that her observations are consistent with latex as a defense, even against monarchs, because sometimes caterpillars simply need to get the latex out of the way, and drinking it may be the only option. Evidence now indicates that sequestration by monarchs is typically not limited by the amount of leaf cardenolides (they choose intermediate cardenolide plants for oviposition and can concentrate cardenolides as caterpillars), so there would be no need to drink latex; in fact, latex is simply not nutritious or beneficial for the caterpillar with all of its concentrated defense chemicals. In addition to the highly viscous and fast-drying nature of latex, it is full of other nasty compounds, including cysteine proteases, which eat away at the caterpillar's gut lining. More latex usually means reduced monarch growth (fig. 5.8).

I feel lucky to have caught that moment of a monarch caterpillar drinking milkweed latex. That a butterfly had happened to lay an egg on a milkweed in our Earth Sciences Center garden in downtown Toronto was unusual. That it survived to become a nearly mature caterpillar had a probability of 1 in a 100. But likely even more important to observing this event was the fact that the milkweed plant was tall, and therefore had two attributes. First, the plant had extra highly pressurized latex to exude, creating a bigger droplet than usual. This giant glob of latex forced the caterpillar to suck it up, because wiping it away presumably did not seem like an option. And the second important attribute of this milkweed was, from my perspective, there was a perfect angle looking up from the picnic bench to observe this brief window into the natural history of coevolution.

DEFENSE, OFFENSE, AND PLANT DEFENSE REDUX

And so a caterpillar has survived the first day of life and a series of barriers to establishment. Trichomes are shaved, latex is disarmed, and the caterpillar is now accessing the more edible portions of a milkweed leaf. Plants make cardenolides as an additional defense, but monarchs are highly insensitive to cardenolides. That is not to say, however, that they are completely insensitive. The previously held paradigm of specialist insects being completely immune to plant defenses like cardenolides is simply not the case. For monarchs and cardenolides, several lines of evidence tell us that caterpillars are still hindered by the toxins, at least under some conditions. First, as was shown physiologically in chapter 3, the extracted sodium pump of monarchs, when placed in test tubes, can be inhibited by cardenolides at high doses (see fig. 3.7). In some field studies, correlations have revealed that caterpillars that feed on milkweed plants with higher cardenolide levels have lower growth than those that eat plants with lower amounts of cardenolides. And when cardenolides are painted on leaves, especially nonpolar cardenolides (those that do not dissolve in water), monarchs often reduce feeding, grow less, or show other signs of stress. But truth be told, the effects of cardenolides on monarchs are quite often difficult to show. In my many years studying monarchs on milkweeds, only a few times has it been clear that cardenolides were negatively affecting caterpillars. The likely scenario is that nutrient limitation or other forms of stress, including the injurious effects of trichomes and latex, ultimately compound to make cardenolides somewhat detrimental to monarch caterpillars.

What happens next, as caterpillars seemingly feed with impunity, makes plants seem sentient in a way that few thought was possible in previous decades. Despite their lack of a central nervous system, legs, or for that matter, "behavior" as defined by many biologists, plants are far from passive organisms. Charles Darwin recognized this, and twenty-one years after the publication of *On the Origin of Species*, Darwin, assisted by his son Francis, published a book titled *The Power of Movement in Plants* (1880). This treatise was an exposition of how plants respond to stimuli, be they light, touch, or gravity. The Darwins developed the argument that plants and animals have similar behaviors, an idea that is currently gaining popularity among plant scientists. Just as the monarch deploys flexible behaviors depending on the attributes of leaves (shaving trichomes or trenching latex canals), so too do milkweeds respond plastically to being fed upon.

Before I go much further in describing how milkweeds respond to the feeding of an established monarch caterpillar, let's look at the history of how plant responses have been studied. Darwin's focus was largely on plant responses to abiotic stimuli (like light and touch, although he did discuss carnivory by Venus flytraps). But it was not until the 1970s that theories about plant responses to insect attack were formulated. Perhaps the time was right to hypothesize that, just as animals have immune responses to infection, so too might plants have responses that increase defense against insect attackers. And although it took more than a decade to catch on as a research topic, beginning in the mid-

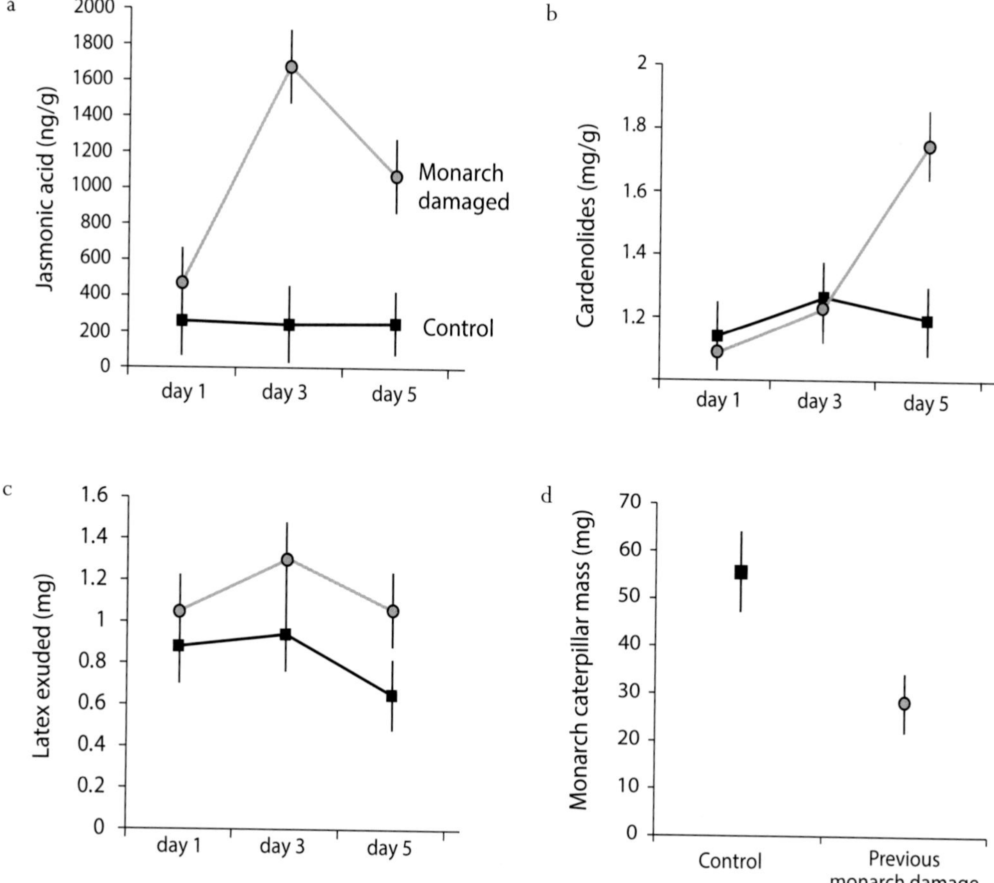

FIGURE 5.9. Monarch caterpillar damage to the common milkweed, *Asclepias syriaca*, causes an immune-like, induced plant defense. Within hours of feeding, (a) the plant shows a burst of jasmonic acid, a universal plant hormone that responds to insect chewing. This response correlates with (b) cardenolide production, which peaks five days after damage and (c) latex exudation, which is enhanced within twenty-four hours and peaks three days after damage starts. (d) When new caterpillars are added to a previously damaged milkweed on day five, they grow 50 percent as much as caterpillars on undamaged control plants. These graphs show the averages of several replicate plants and standard errors. Note that five days of feeding by newly hatched monarchs results in a trivial amount of damage, typically less than 3 percent of the leaves eaten, so it takes very little herbivory to initiate the induced defense response.

1980s, it was becoming clear that most, if not all plants respond to insect attack with defensive countermeasures. By the mid-1990s, the seminal book on plant responses to herbivory was published by my PhD adviser, Rick Karban. At the time, immersed in Karban's laboratory as a graduate student, I latched on to this topic, termed "induced defense," and conducted my thesis research on the fitness consequences of such responses in wild radish plants. Across essentially all plants, only a handful of hormones are employed to orchestrate all that a plant does. And one of those hormones, jasmonic acid, is primarily engaged in regulating defensive reactions.

Returning to milkweed's defense: After a day of monarch feeding, jasmonic acid rises within plants and initiates the production of enhanced latex and cardenolides (fig. 5.9). What this means for monarchs is that as an individual feeds on a leaf, the plant is dynamically responding with stress hormones, and as such, the plant is increasing defense production. When newly hatched caterpillars are reared on "induced plants," those that have experienced a few days of previous feeding, these new caterpillars grow dramatically less well compared with caterpillars on "uninduced" plants (those with no previous damage, fig. 5.9d). The key point is that plants regulate their investment in defenses in real time, and milkweeds have immune-like reactions to monarch feeding that reduce future monarch performance.

PLANT DEFENSE SYNDROMES

How are we to understand and piece together the multiple facets of milkweed defense, from trichomes, latex, and cardenolides, to the orchestration of their inducibility following attack? One approach, which I have termed the defense syndromes hypothesis, is to start with the observation that most plants deploy multiple defensive traits. Then as we watch the natural history of monarchs on milkweed unfold, it is clear that a series of barriers to feeding are presented by the plant, and that they are progressively overcome or otherwise dealt with by the insect. From the insects' perspective, any one of these traits may not be too problematic, but together they comprise quite a set of challenges. As if the trichomes were not enough, a pulse of latex begins gluing the young and vulnerable caterpillar, and the poisons within the plant add yet another layer of defense. It is perhaps surprising, then, that we know so little about how multiple traits interact to provide a package of defense. In chapter 3, I discussed the diversity of chemical variants produced by a plant in one class of chemical defenses (the cardenolides); there too, I hypothesized that multiple compounds may be more effective than one alone. Accordingly, defense syndromes can be thought of as the plants' version of a one-two punch. Like a cocktail of drugs to treat human disease, the most successful plant defenses act from multiple angles.

The concept of defense syndromes is useful because it allows us to cope with the complexity of myriad defenses and also because it compels us to synthesize the many ways and combinations in which plant defenses are deployed. The one-two-punch likely evolved to solve the following problem: What is the most economical way to deploy a set of defensive traits that are effective? The economical part of this equation is important, as most organisms live under the constraint of limited resources, and thus allocation to defense (or anything else for that matter) should be conservative. Although some plant defenses may be redundant, like a secondary barrier in case the first one fails, they are more likely to be stacked on top of each other such that the combination of defenses is more effective than any one alone. The arms race not only leads to more defenses, but combinations of defenses that are synergistic.

Defense syndromes suggest which plant defensive traits complement each other and have stood the test of time (by this I mean the deep evolutionary time of millions of years). In addition, which sets of traits have repeatedly evolved together, convergently, in distinct plant lineages? The concept of convergence, or repeated independent evolution, is central to understanding what is important in biology, and I will again return to this issue in chapter 7. Looking into evolutionary history is a way to ask the plant, What has been important in your biology? It is a way to be true to our goal in science of seeking general and important patterns, because the plants can show us that through their history, certain traits or associations have repeatedly evolved. For milkweeds, when we first began to look for defense syndromes, it became clear very quickly that one association is very strong across species, irrespective of their relatedness: latex exudation and trichome density are positively correlated (fig. 5.10). That is, on average, milkweed species with few trichomes, for example *A. incarnata* and *A. exaltata*, exude dramatically less latex than species with densely trichomed leaves, such as *A. californica* or *A. lemmonii* (the latter of which holds the record for exuding the most latex of any *Asclepias* species). Common milkweed, *A. syriaca* is somewhere in the middle.

The cause of the positive latex-trichome relationship is unclear, but it is certainly worth discussing, as the pattern is quite strong. The first hint about the function of this relationship came from plants that break the pattern: the exceptions that prove the rule. Among the milkweed species that do not follow the pattern are the several that have a wax layer covering their leaves (see fig. 5.10). As milkweed leaves that have wax have essentially no trichomes, we might expect these species to be low in latex; but on the contrary, they are up there among the top 15 percent of latex producers, plants that we would expect to be quite hairy. As discussed at the beginning of this chapter, the physiological function of waxes is highly similar to that of trichomes, and they also block caterpillar access to leaf tissue. High latex content in waxy species is consistent with a link between leaf covering (not trichomes per se) and latex production.

The evolutionary connection between leaf covering and latex may have to do with the hot and arid environments in which most milkweeds grow, perhaps because latex is especially effective against herbivores when they are already dealing with trichomes on their first day of life. In other words, in an evolutionary sense, it is plausible that leaf coverings evolved in response to heat and aridity but also provided a very useful barrier to feeding. A caterpillar that has to suffer through mowing the "leaf lawn" may be especially susceptible to a pulse of thick, sticky, and toxic latex. It may also be that in such hot and

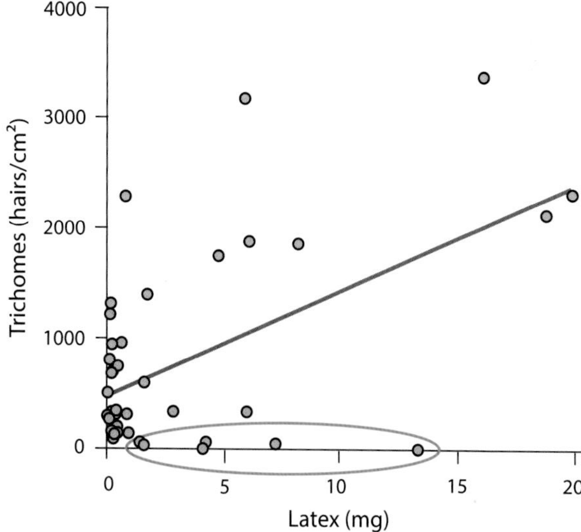

FIGURE 5.10. Latex exudation and trichome densities from forty-five *Asclepias* species. Each dot represents the mean of several replicates for a distinct species. Note the species that are circled; these are milkweed species with a layer of wax crystals on the leaves (and no trichomes). Nonetheless, these species exhibit substantial latex exudation. As discussed in the text, they are the exceptions that prove the rule, because the waxy leaf covering is physiologically similar to trichomes. On average, plants present latex and trichomes together. Where this is not the case, milkweed species tend to have a thick layer of wax on their leaves.

arid environments, the relative construction costs of trichomes (or wax) and latex are economically viable. Under other conditions, particular resources may be more plentiful, and hence, other defensive strategies may be favored.

As a final note about defense syndromes, and perhaps what convinces me the most about the joint evolution of trichomes and latex as a strategy against herbivores, is another exception that appears to prove the rule. The trichome-latex association across species need not be this way. What I mean to say is that trichomes and latex are not actually connected in a physiological or genetic sense. Yes, they are both largely composed of carbon-based building blocks, but little else suggests that if a plant increased production of one, the other should necessarily follow. As a case in point, when we have repeatedly examined distinct clones of common milkweed, either from a single population, or across its geographic range, we find tremendous variation in both the density of trichomes on leaves and the amount of latex that is exuded upon leaf damage. But the two traits are not correlated within common milkweed: genotypes can be found that produce high trichome densities but either high or low amounts of latex. Therefore, as natural selection operates within a species such as *A. syriaca*, trichomes and latex do not necessarily evolve together. This suggests no physiological or genetic connection between the two traits, and it implies that the association between them across different milkweed species is the result of adaptation toward a particular strategy. Among milkweed species, trichomes and latex form the backbone of a defensive syndrome that has repeatedly evolved.

WHAT ELSE IS A PLANT TO DO?

The arms race seems relentless, an ever-increasing investment in defenses and counterdefenses. And yet, in a coevolutionary sense, the arms race may not be the most stable strategy, as organisms should eventually run out of resources or effective strategies. It could end in stalemate. But interestingly, plants defend themselves in other ways, which do not seem to engage in the same escalating arms race. Take for example the "tolerance" or regrowth strategy. As milkweeds store more of their resources underground in their roots, they are more capable of regrowing following defoliation by monarchs. Such a tolerance strategy, living with its herbivores so to speak, has been favored in the most recently evolved milkweed species. That is not to say that tolerance is independent of an ongoing arms race, but it is part of the complicated mix of many traits involved in the plant-herbivore interaction.

Another important piece to the dizzying puzzle of so many plant defenses involves the enemies of the monarch. As we will learn about in greater detail in the next chapter, not only birds, but many other predators, including wasps, flies, spiders, and stinkbugs, are happy to make a meal out of monarchs, especially when they are caterpillars. From milkweed's perspective, "the enemy of its enemy is its friend." This concept, which traces back to an East Indian treatise on military strategy written more than two thousand years ago, simply suggests that plants ought to employ, encourage, or otherwise engage with the enemies of monarchs. What a clever plant defense, calling in predators! And in fact, when milkweeds are damaged by monarch caterpillars, they release a blend of volatile organic compounds that are attractive to wasps that parasitize monarchs. This type of interaction, plants signaling to get a little help from their friends, is exceedingly common in nature and has been studied in dozens of plant species.

Like tolerance of herbivory, biologists believe that calling in predators of insect herbivores is a more stable (less escalating) strategy than to develop traits typically involved in a coevolutionary arms race. Why? Because here the plant is engaging a third party, so if there is natural selection on the monarch to reduce these volatile plant signals, it has to evolutionarily battle both the plant and the predators, as it is in both of their interests to conspire against the monarch. There is simply less the monarch can do. Yet again, tolerating herbivory or signaling to predators is just one part of the milkweed's defense repertoire.

Before I end this discussion, a brief disclaimer about how we study and think about coevolution: Simply by observing the species in their current state of interactions, we are unable to really pinpoint the evolutionary origin of the traits and behaviors we observe. In this chapter we discovered several important characteristics of milkweeds, including trichomes, latex, and cardenolides. Even though these traits are all important defenses for the plants and evolved as part of an arms race, we could not have concluded this from observations alone. Each of these traits had a different history of natural selection that resulted in its current existence. Our studies on the evolution of milkweeds have revealed this varied history, and while latex and cardenolides seem to be coevolutionary traits, the evolution of trichomes was decidedly more complex. Trichomes were evolutionarily coopted as a defense after they had fulfilled their original role in protecting leaves from heat and aridity. Trichomes are surely a barrier to caterpillars and have undoubtedly imposed natural selection on monarchs to adjust their feeding behaviors, but it is unlikely that the origin of trichomes had anything to do with insects. Other traits, like latex, have no known function other than defense. But even so, whether latex arose evolutionarily in response to grazing by mammals, specialized insects (only one of which is the monarch), or perhaps even in response to microbial pathogens, which are flushed out of wounds by latex, is unknown. Finally, the sequence in which the natural history plays out as we watch it may not reflect the evolutionary history in which the traits evolved. In other words, the observed sequence of the interaction does not necessarily recapitulate the evolutionary sequence of the interaction.

Despite these cautionary notes, the weight of the evidence from our historical investigations using phylogenies and studying natural selection in the wild points to a coevolutionary arms race between monarchs and milkweed. Inducible plant defenses are like an immune system. And the combination of the multiple defenses provides us with the defense syndromes hypothesis for why defenses may be deployed simultaneously. Further investigation will surely unravel many more mysteries, far stranger than science fiction, and yet the result of a natural, escalating arms race over the millennia.

Having survived the precious first few days of life, munching away on milkweed, the monarch caterpillar now enters the exponential growth phase. It has several simple and straightforward goals. Eat, eat, and eat more milkweed. But in addition, it also must effectively sequester away milkweed's toxins and avoid being eating by predators.

CHAPTER 6

Saving Up to Raise a Family

Things aren't always what they aposeme:
The model monarch as a caterpillar eats
Milkweeds, stores their hearty poisons,
and the butterfly defeats
Some portion of an avian predation team
By advertising—with show of color, lack of haste—
That those that dare to peck will find it in bad taste;
But the monarch straddles defence with another gimmick
When it feeds on milkweed species in which there are
No poisons and becomes an edible adult,
the so-called "automimic."
(Or is that a chrysomelid beetle that recalls
the German peoplescar?)

—John M. Burns, "Up the Food Chain"

Eat, rest and digest, poop, hide. Repeat. Eat, rest and digest, poop, hide. Repeat. Monarch caterpillars are extraordinary eating and growing machines. They hatch from an egg that weighs as much as tiny bread crumb, about half a milligram. Through childhood, caterpillars grow to well more than two thousand times that mass. If an eight-pound human baby were to grow at this rate, by the end of one year, she would weigh as much as thirty-five of the largest adult elephants, about half a million pounds (more than 220,000 kg). But after ten to twelve days, monarch larvae won't be hungry or little anymore; they will be big, fat caterpillars. Eating leaves is no simple way to make a living: they plow through a lot of leaves, lots of roughage, with very little protein to build a body. Leaves are not nutritionally rich or well-balanced for an animal like the monarch, which must grow fast, develop strong wing muscles, and avoid being eaten along the way. Then there are the poisons. Monarchs go to great lengths to not let them touch their internal organs, and yet some are stored away to advertise to enemies that they are distasteful. Monarch caterpillars eat and try to avoid being eaten. Yet only a small fraction of their brethren survive their first day of life. Milkweed, with its trichomes and latex barriers, is not an easy food to eat. On top of that, the plants emit odors reflecting their poor quality to adult butterflies and also release volatiles that send signals to insect-hunting predators. In the end, very few monarchs ever establish themselves as feeding machines.

The few monarchs that survive the first few days now need to grow. As explained in chapter 2, to eat and avoid being eaten are two of the most important things any organism does, and all in the service of survival and reproduction. Although a common definition of any "adult" animal rests on its reproductive capability, in the Lepidoptera (in fact, also flies, beetles, and wasps), reproduction is left to a discrete stage after metamorphosis (into an adult butterfly or moth). The caterpillar itself does not bother with this other important life goal. In addition, caterpillars typically do not even choose where to eat, since adult butterflies are much more mobile, and they choose where eggs are laid. The hatched caterpillar is thus an eating machine (fig. 6.1), unfettered by decisions about what food to eat or where or when to reproduce.

Monarchs (and a few other Lepidoptera) are unusual in an important respect compared with most other butterflies and moths, and perhaps the great majority of other organisms. They do not seem to experience the nearly universal trade-off between growing fast and avoiding predators. The growth stage, although critical for all organisms to gain the resources they need to mature, is typically fraught with danger, and the faster an organism grows, the more susceptible to some dangers it often becomes. In other words, the relationship between "eat" and "don't be eaten" is atypical for monarchs. Let me explain.

a

b

FIGURE 6.1. Like all insects that undergo metamorphosis, monarchs have compartmentalized their growth versus reproductive functions. (a) The caterpillar is specialized as an eating machine, largely a tube filled with ground milkweed leaves. (b) The bulk of the leaf material is not digested but rather expelled as poop (in insects, technically called "frass"). The adult butterfly is a reproductive and flying machine. In fact, adults do not grow or eat, except to drink water and floral nectar as fuel.

Fast growth depends on eating, and eating can be dangerous. Elizabeth Bernays, retired professor from the University of Arizona, having watched individuals of a caterpillar species for 3,800 hours in the field, concluded that feeding was their most dangerous activity. Although caterpillars spent only about 3 percent of their time feeding, this was when 80 percent of the mortality by predators occurred. This pattern may be generated by many factors, but most likely the act of feeding simply offers visible, audible, and odoriphorous cues to would-be predators. Consequently, many caterpillars spend most of their time inactive, quietly hiding, and perhaps visually blending in. They may also feed at night or otherwise conceal themselves because feeding is dangerous. For most caterpillars, the more they eat, the more they are exposed to predators. However, there is nothing cryptic about monarch caterpillars, not how they look, where they feed, how they feed, or when they feed. This too is elaborated on in this chapter, as we examine the impact of bright, contrasting coloration, mimicry, and predation.

ON BEING A VEGETARIAN

Leaves are not a great way to build a body. Take protein, which is one of the most important building blocks of an animal. Protein is largely composed of amino acids, which are typically limited by the availability of nitrogen. An average milkweed leaf is about 3 percent nitrogen (on a dry mass basis), but an average animal body is about 10 percent nitrogen. This mismatch means that herbivores' growth is often limited by lack of nitrogen, as they have to process a large amount of plant material to build their bodies. It perhaps seems odd that nitrogen is limiting, given that nitrogen is everywhere, and nearly 80 percent of our atmosphere is made of gaseous nitrogen (N_2). But this form is not available for use as a biological building block to most plants and animals because the two nitrogen atoms are so strongly bonded. The strong bond makes N_2 largely inert, or hard to break apart to use for other purposes. Accordingly, plants gain the bulk of their nitrogen from nitrate (NO_3^-) and ammonia (NH_3) in the soil, two forms of nitrogen that are contributed by microbes and more easily used to build chlorophyll, proteins, and DNA. However, while processing large amounts of leaves to concentrate this nitrogen, herbivores encounter defensive traits of plants, in addition to vast quantities of undigestible fiber, which is made mostly of cellulose.

As a consequence of this diet, herbivores are often slow and large, and use various strategies (high throughput or slow fermentation) to process leaves.

The herbivorous lifestyle has its benefits, the most notable of which is having a sessile and sustainable food source, making food relatively easy to find. In fact, this argument has been used to advocate for vegetarianism in humans. Plants primarily require sunlight, carbon dioxide, and water to grow. And given the tremendous losses that occur each time energy moves up a link in the food chain (only about 10 percent of the energy at one trophic level is turned into mass at the next higher level), eating closer to the base of the food chain (plants) is less wasteful of energy. A salmon, for example, feeds at the third or fourth trophic level, high on the food chain, primarily eating other fish. As such, a tremendous amount of energy was lost in the conversion from algae to herbivorous fish, and then again in the conversion to small carnivorous fish, and so on, as we move up the food chain until humans eat the salmon. In fact, when we eat that salmon, probably greater than one hundred times more biological energy has gone into making that meat than its caloric content. Herbivores are the original converters of sustainable energy into meat.

However many the benefits of eating leaves, it is hard to argue with the fact that leaves are a relatively dilute and well-defended form of food. Meat, being much more concentrated in protein and fat, allows organisms to take fewer meals and longer between them. As predators, cheetahs, snakes, and spiders each eat prey that are "concentrated foods" rich in resources. And these carnivores can often drag or otherwise squirrel away their meal out of reach from others. Herbivory is not only one of the least efficient ways of gaining nutrition, but herbivores are often big, slow, and exposed compared with their predaceous counterparts.

All this puts herbivores, like caterpillars, between a rock and a hard place. In a trophic sandwich as it were, between plants that are well-defended and carnivores that want to eat them. On the one hand, herbivorous insects eat a toxic, nutrient-poor food, and on the other hand, they are often subject to intense predation pressure by carnivores. The monarch butterfly, however, has found an escape from this apparent conundrum. Monarchs use that poor food as a defense, and they advertise it.

WARNING: DON'T EAT ME

Herbivory, the act of consuming plant tissues, is usually not a life-or-death situation for the plants being eaten. In fact, it is relatively rare, even when herbivorous pests defoliate a plant, for that plant to actually die. Most plants resprout new stems, re-flush leaves, or simply wait until next year. In contrast, consumption of animals by carnivores involves death. As such, there is tremendous natural selection for animals to resist being eaten, and animals go to great lengths to defend themselves. Escape behaviors (such as fast running), hiding, blending in (crypsis), and feeding at a time when predators are not active are perhaps the most common features that animals use to avoid predation. But monarch caterpillars do none of that. They feed in the open, on the tops of leaves during the day, and stand out on a green background with their highly contrasting white, yellow, and black stripes. These markings are thought to be aposematic (or "warning!") coloration, a cue for predators to not eat them because they have sequestered away plant poisons.

Over the course of about ten years, three contemporaries of Darwin articulated a role for plants in the toxicity of brightly colored butterflies. First, in 1887, J. W. Slater noted that "strikingly-coloured insects, not otherwise specially protected, will be found to feed upon poisonous plants, or upon such as, though not poisonous, possess unpleasant, or at least very powerful, odours or flavours. From such a diet I conceive that the insects in question may receive properties positively injurious, or at least, disgusting, to their enemies, and that a brilliant colouring may therefore here serve as a danger signal, like a quarantine flag, warning all comers to keep their distance." Alfred Russel Wallace, one of the great scientific intellectuals of the nineteenth century, and codiscoverer of evolution by natural selection, had a fondness and deep understanding of butterflies. In his 1889 book, *Darwinism*, he penned ideas similar to Slater's about the origins of distastefulness of some butterflies: "[Heliconidae] owing to the food of the larvae or some other cause, possessed disagreeable juices that caused them to be disliked by the usual enemies of their kind."

Four years later, the German lepidopterist Erich Haase published his important book *Researches on Mimicry*, in which he further developed this idea and introduced the term "pharmacophagy" (derived from the Greek roots *pharma* = "poison" and *phagy* = "eat"): the act of herbivores consuming toxic plants and accumulating toxins within their bodies for their own defense against predators. Although Haase developed this hypothesis in view of his beloved swallowtail butterflies (Papilionidae), the milkweed butterflies played an important role as supporting evidence. Haase wrote, "It is without doubt chiefly the peculiar food of the larvae, which, through the storage of certain, mostly emetic poisons, renders first of all the larvae and then the pupa and even the imago itself very unpalatable if not actually injurious morsels." Later on the same page Haase refers to the purgative properties of milkweeds, which are fed upon by monarchs.

Despite milkweed's toxicity and monarch's aposematic coloration, walk through any field of milkweed with abundant monarchs, and you will see evidence of predation on the caterpillars by all kinds of invertebrates, including spiders, ants, wasps, and stinkbugs (fig. 6.2). Predation rates by these invertebrate predators have been estimated to be greater than 95 percent in some populations. Such a high risk of predation after surviving on such well-defended plants! How and why is the monarch's aposematic coloration failing?

Perhaps aposematic coloration is effective against vertebrate predators. Birds are well-known consumers of caterpillars. When birds are excluded from plant canopies, for example, caterpillar abundance typically increases by more than 100 percent. In a classic study, Bernd Heinrich, a well-known natural history writer and now emeritus professor at the University of Vermont, patiently observed caterpillars of several species in the field for many hours. He reported that monarchs were unusual in how they fed in the open, during day and night, and yet were seemingly avoided by bird predators. Nonetheless, predation rates on other caterpillars he observed, often cryptic species, were quite high.

Twenty-five years later, experiments isolated the cause of Heinrich's observations. A recent doctoral student, Colleen Hitchcock, took an elegant ap-

a

b

c

FIGURE 6.2. Natural predation of monarch caterpillars. (a) Note the black and yellow warning coloration of the vespid wasp. These wasps often "gut" monarchs, leaving behind the food bolus and carrying off their meat. (b) A stinkbug carries off this dead and decaying monarch, a meal for later. (c) Parasitoids, which develop inside the host caterpillar, eventually come out to pupate (shown here are three *Lespesia archippivora* fly larvae that emerged from a single monarch).

proach to studying aposematic coloration and its effects for caterpillars. Hitchcock took all of the other things that make caterpillars "caterpillars" out of the study and focused solely on coloration. Great science often involves such isolation of a single factor and studying its effects. She made more than three thousand model clay caterpillars, constructed in the likeness of aposematic monarchs and also of more cryptic, solidly colored green or brown forms. By putting these otherwise identical clay caterpillars out in the field and observing peck marks on the clay models, Hitchcock concluded that bird predation attempts were high on the monarch models (21 percent) but were disproportionally higher and quicker on the cryptic forms. So, birds more strongly avoid aposematically colored caterpillars, despite being easier to find than cryptic species. Nonetheless, a high diversity of invertebrate predators do not seem to care about this aposematic coloration. Why?

SEQUESTRATION AND COLORATION

The theory behind aposematic coloration is that a visual cue to potential predators is information to avoid toxic prey. As such, the warning may be considered a lifesaver for the prey. After all, a dead caterpillar that a predator spits out because it is noxious is still dead. And perhaps the dangerously colored animal is also doing the predator a favor: don't waste your time and risk getting ill. In this section, I develop a hypothesis for why warning coloration is a meaningful and effective way of communicating with vertebrate predators like birds, but why it fails with invertebrate predators. The hypothesis has to do with the specific cardenolides monarchs sequester and their impacts on vertebrate versus invertebrate predators.

To start constructing the hypothesis, we must return to cardenolides, chemical polarity, and toxicity. Recall that even monarchs are somewhat sensitive to high doses of specific cardenolides. Most milkweed species, including the common milkweed, *A. syriaca*, contain a multitude of different cardenolide forms ranging from the polar (highly water soluble) to the nonpolar (primarily soluble in fats or lipids). Although we do not specifically understand the consequences of producing such diversity in cardenolides, it is clear that the different compounds have differential effects on animals. And across the board, it seems that nonpolar cardenolides are the most toxic. This high toxicity appears to be a consequence both of these compounds' ability to freely cross lipid membranes of animal cells (getting to the target sodium pump) and also of stronger bonding once cardenolides arrive at the sodium pump. Thus, despite the monarch's relative insensitivity to cardenolides, it still may be in its interest to specifically sequester more polar cardenolides that are the easiest to store and least toxic to the monarch's own body. Indeed, this is what research has shown. Monarchs preferentially sequester highly polar to intermediate polarity cardenolides over nonpolar forms.

Why monarchs differentially sequester polar cardenolides is not completely understood, although reduced risk of self-poisoning certainly stands to reason. How exactly they sequester the different cardenolides is not completely worked out either, but it is clear that the monarch caterpillar employs physiological mechanisms that actively transport the compounds. For example, there is evidence that physiological processes move some of the cardenolides into the body, that other cardenolides are prevented from passively diffusing into the body (and are excreted), and that still other cardenolides are biochemically transformed into other forms. The bottom line, however, is that monarchs sequester disproportionally more polar cardenolides than are in their host plants. As we will see below, and in greater detail in chapter 8, one reason that many birds are strongly affected by cardenolides is that they have a strong sense of taste and a highly sensitive form of the sodium pump.

Although different cardenolides have different potential to induce vomiting in birds, even highly polar compounds cause some nausea. Additionally, because of their large size and lack of dexterity, most birds consume whole caterpillars. Meanwhile, because of their small size, invertebrates can chop monarchs into pieces to selectively feed on their least defended parts (see fig. 6.2). Importantly, Lincoln Brower showed that birds are able to taste cardenolides (when isolated compounds were added to artificial food) and reject cardenolide-dosed food even at concentrations substantially lower than that which causes vomiting. Consequently, aposematic coloration may be a very effective way for caterpillars to communicate with bird predators and, ultimately, to avoid being eaten.

Relevant to our understanding of sequestration and warning coloration is how these traits impact other enemies, such as parasitoids, which can have dramatic effects on monarch populations. This term, parasitoid, refers to a seemingly intermediate strategy between predators (like the stinkbug in fig 6.2) and parasites (like the protozoans described below). A parasitoid has several distinguishing characteristics. First, most parasitoids lay their eggs on top of or inside (by injection) a host larva. Second, parasitoid larvae (parasitoids are typically flies or wasps) eat the host organism from the inside out. And third, although parasitoid development nearly always results in the host's death, it is not immediate, as it is in the case of predators. Instead, caterpillar death is typically delayed until the caterpillar (and parasitoid) has nearly completed development.

What is surprising about monarchs is that they are attacked by so many species of parasitoids, including several generalists that will attack a diversity of other caterpillars. Under some conditions, rates of attack by parasitoids in the field can be 100 percent. That is, occasionally every monarch caterpillar in a particular location, although feeding away, harbors a parasitoid inside and will never live to become a reproductive butterfly. It also seems that these parasitoids are undeterred and apparently unaffected by sequestered cardenolides. Perhaps because parasitoid larvae are free-moving inside the monarch body, they can avoid the most toxic parts?

When under attack by invertebrate predators or parasitoids, monarchs are not defenseless. Caterpillars respond to sounds, especially those in the frequencies created by flying invertebrates. Monarchs perceive sounds by sensory sensilla along the outside of their bodies. And in response, monarch caterpillars have many behaviors, most prominently biting and vigorously flicking their bodies. Such flicking is common in lepidopteran larvae, and has been shown to be effective at reducing attack, especially by flying predators and parasitoids. Given that their sequestered chemical defenses are apparently ineffective against invertebrates, monarchs seem to rely on detection of flying predators and parasitoids and attempt to shoo them away physically.

In sum, not only can birds see a monarch caterpillar's aposematic coloration, they can taste cardenolides, which are toxic to them, and this reinforces avoidance of these bitter caterpillars via their warning coloration. It is far less clear whether invertebrate predators see aposematic coloration, taste the bitterness of cardenolides, or experience the toxic effects as strongly. As such, mortality remains high despite the bright coloration of monarchs, and survival as a caterpillar is surely not easy. Finally, there are the debilitating microparasites of monarchs. These are not visually oriented enemies (and it is difficult to see them too), but these parasites appear to be influenced by sequestered plant defenses. A surely productive area of future research would be to contrast the effects of warning coloration and sequestered toxins on vertebrate predators, invertebrate predators, and microbial parasites.

MONARCHS USING MEDICATION

So far, I have talked extensively about how plants, insects, and birds are intensely chemical organisms. And although we may not often be aware of it, so too are we humans. Think of our pharmacopeia, the vast quantities of medicines we consume. Some 50 percent of our pharmaceuticals are derived from nature, half of which are derived from plant extracts. Insect sequestration of plant compounds is an analogous process, a potential medicine against parasites and predators. As discussed in chapter 4, adult female monarchs make decisions about where to lay eggs in part based on the cardenolide content of milkweed leaves. When caterpillars hatch out on higher-cardenolide leaves, they tend to sequester more of the toxin. Hence, behavioral decisions by adults affect the level of sequestration by their offspring.

A major discovery was made in 2010 that has changed the way scientists think about the extent to which animals, especially monarchs, use plant compounds to medicate themselves and their offspring. Jaap de Roode and his laboratory at Emory University have demonstrated that parasite-infected adult monarchs medicate their offspring by changing butterfly oviposition preferences. What is most interesting about this interaction is that neither adult butterflies nor juvenile caterpillars are able to "self-medicate," but rather, infected adults make oviposition choices that medicate and maximize the fitness of their caterpillar offspring.

Monarchs are highly susceptible to a tiny protozoan parasite, *Ophryocystis elektroscirrha* (known as "OE" among monarch biologists, for the obvious reason that nobody can pronounce the Latin name). OE has a remarkable ability to infect monarchs and can reduce their ability to produce young. This parasite typically does not kill the butterfly (fig. 6.3). Nonetheless, given the intimate parasitic association, it should not be surprising that natural selection would favor antiparasitic strategies in monarchs. Adult butterflies apparently sense that they are infected and that their offspring will likely be infected as well, and accordingly prefer to lay eggs on a medicinal milkweed. Infected monarchs shift their preference toward *Asclepias curassavica*, which is higher in cardenolide concentrations than the other milkweeds tested. A higher cardenolide food means greater sequestration of cardenolides, thereby reducing the spore load in the larvae. The critical test for such medication is not only that the animal ingests some therapeutic food, but that this food is typically less desirable and yet becomes preferable (and beneficial) upon parasite infection. In other words, the hallmark of medication is an adaptive behavioral switch initiated by being sick.

Despite this spectacular discovery, several unanswered questions remain. For example, why don't monarchs always show a preference for *Asclepias curassavica*? Are there costs associated with eating this plant when monarchs are not infected? Are there properties of this plant, in addition to the cardenolides, that are antiparasitic? As we learned in the previous chapter, milkweed latex is a cornucopia of biologically active chemicals, and some of these compounds may be critical for the medicinal effect. If cardenolides are indeed the causal agent of resistance, which specific compounds are antiparasitic (and why)? Whatever the case may be, monarch caterpillars themselves are not known to be self-medicators; it is only the mom that can impose this treat-

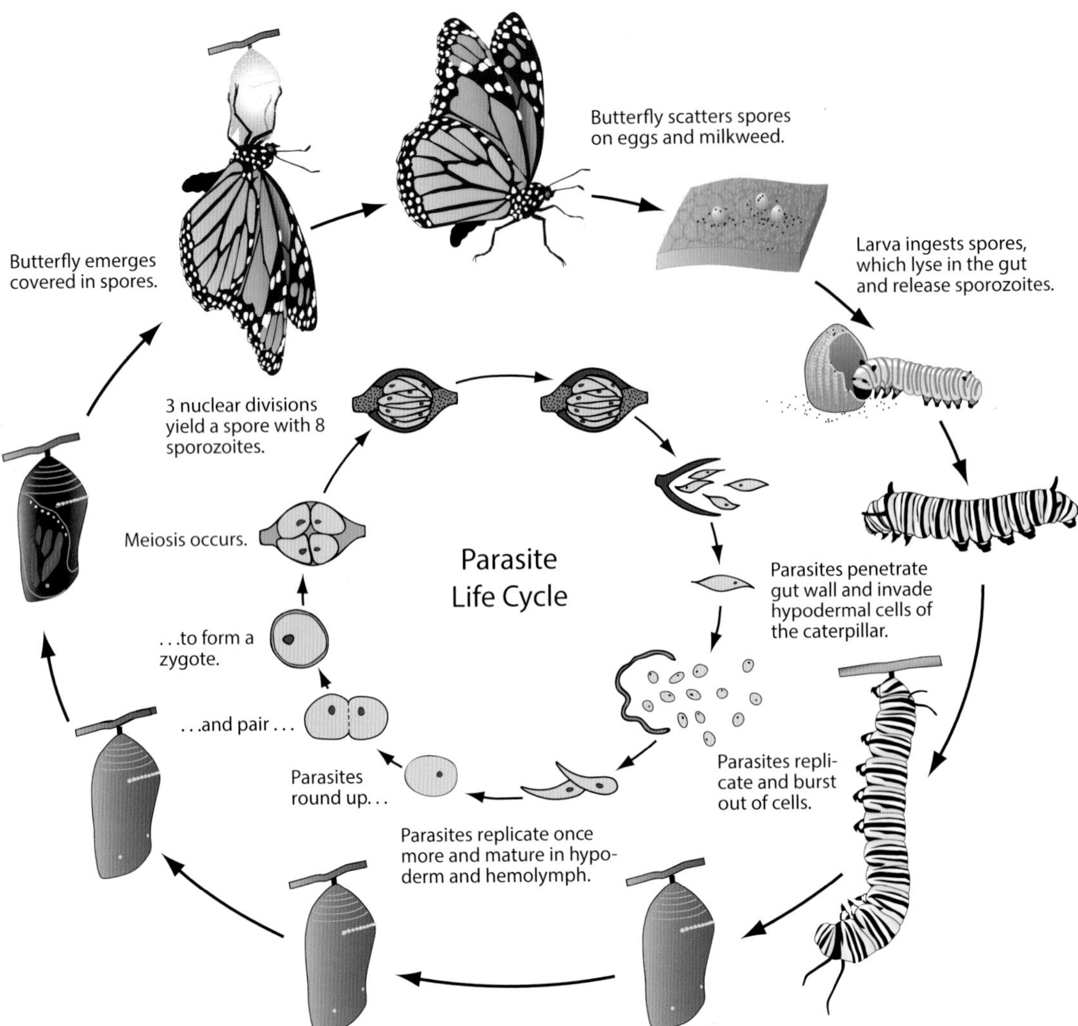

FIGURE 6.3. The parallel life cycles of the monarch butterfly and its highly specialized protozoan parasite, *Ophryocystis elektroscirrha*, aka "OE." Organisms are not drawn to scale.

ment through her oviposition decisions. When on leaves with higher cardenolide levels, greater sequestration by larvae serves not only as preventative medicine by warding off bird predators, but also as a direct medicine against microbial parasites.

THE WANDERING PHASE

Should a caterpillar be so lucky to be part of the less than 10 percent that make it to full size, its "growth phase" is done. Having accumulated more than two thousand times the mass it started with, it is time for the monarch to achieve some fitness via reproduction. All caterpillars, whether found in the spring in the southern United States, or in the summer throughout the breeding grounds in the eastern United States and Canada, are on a mission to reproduce. To accomplish this, metamorphosis into an adult butterfly is the next critical step. In all insects that undergo complete metamorphosis (Lepidoptera, beetles, flies, and wasps), they transform from wormy larvae to adults in what is called a pupa. For butterflies, in particular, a special term is used for the pupa: "chrysalis." But before a caterpillar pupates, it first empties its gut, removing all food waste. Commonly, the last bits of frass (insect poop) excreted have a pink or reddish color.

Most Lepidoptera also have a wandering phase before pupation. For a monarch, this can last several hours and often involves leaving its natal plant (or patch of plants). The caterpillars are no longer interested in food; they often move quickly and do not grip tightly on to things (for example, leaves, twigs, and human fingers), as they would before this phase. Entomologists believe that the wandering phase represents a time to get away from "the evidence"—leaf damage, frass, and other cues that may lead a predator to finding the chrysalis—and to find an ideal place to hang and transform over the next one to two weeks. This is usually a sheltered place, sometimes under a milkweed leaf, but more typically on a railing, fence, tree branch, or some other crevice with a roof. The chrysalis is not only a critical transformation phase but one of near complete physical inactivity, where the animal has little potential to defend itself behaviorally (fig. 6.4). As such, the wandering phase is important for finding a quiet and protected spot.

Perhaps the most striking features of the monarch's chrysalis are the gold spots (fig. 6.5). Innumerable children, scientists, and non-scientists alike have

FIGURE 6.4. The monarch's pupation sequence, with this transformation occurring over about thirty minutes. The hardening process including the setting of glistening gold spots takes several additional hours.

stared at the monarch's chrysalis and have wondered about those dots of gold. The jade dumpling, beautiful as it is, decorated in this manner becomes a gilded lily. Do the gold and black on the chrysalis impart warning coloration, perhaps startling a predator passing by with bright, shiny flecks? In my own musings about the monarch chrysalis, they have never seemed aposematic; they just seemed ornate, which is not something natural selection usually favors for the fun of it. But a few summers ago, my thinking about these gold-spotted pupae took on a different perspective. I hired a former undergraduate in my laboratory, Ellen Woods, who is a gifted plant ecologist and nature photographer. In fact, many of her photographs grace the pages of this book. She was between finishing her degree at Cornell and starting a graduate program at Wesleyan University, and I asked her to take pictures of monarchs and milkweed. The images of pupae that we typically see accentuate the ornate gold and black on jade (as in fig. 1.1b). But what does a potential predator see? When Ellen got under a milkweed plant in my front yard, her pictures spoke a thousand words. It is now clear to me how these flecks, and the gold ring, can blend

FIGURE 6.4. (*continued*)

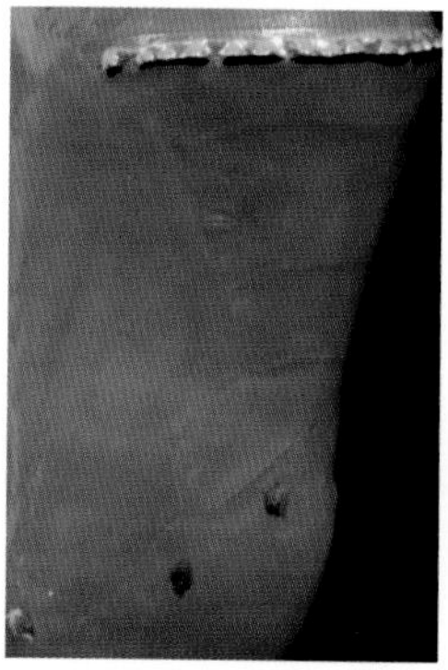

a

FIGURE 6.5. (a) The pupal stage (or chrysalis) of the monarch butterfly with its gold spots. (b) and (c) The chrysalises are shown in two natural poses on milkweed, where they seem to blend in (see also fig. 1.1).

b

c

in and provide crypsis. The quiescent pupal stage of the monarch, unlike the more mobile caterpillar and adult stage, melds into the background.

As monarchs transition from feeding machine to quiet, metamorphosing jewel, they also transition from warning coloration to camouflage. Although

FIGURE 6.6. The "eclosion" sequence, photographed over a period of ten minutes. Eclosion is the entomological term for adult emergence from the pupa.

sequestration is common to both stages, the apparent lack of mobility or behavior has favored a cryptic visual strategy in the chrysalis stage over the ostentatious coloration as a caterpillar. And such is the nature of natural selection: optimization given the materials and conditions at hand, but not always a linear path. As adults emerge from their chrysalises, yet another strategy surfaces (fig. 6.6).

THE IMITATION GAME

Mimicry is one of those phenomena that, when you see it, brilliantly reinforces the terrific power and seeming magic of evolution by natural selection. Mimicry is the similarity, usually in appearance, of species that inhabit the same region, even though they are not closely related. In other words, it is the evolutionary convergence on form despite the lack of recent shared ancestry. Adult monarchs are aposematically colored, with their orange, black, and white patterning, and have packed away toxic cardenolides from milkweed. As such, they are toxic "models" for other species to evolve to

FIGURE 6.6. (*continued*)

mimic because of their successful and honest strategy of coupling visual appearance with toxicity.

In some of the most extreme cases of mimicry, for example, those involving coral snakes, a single nontoxic snake species may evolve to look like different toxic snakes in different geographical locations. What makes this such a convincing case of adaptation is that a single species is expected to have a uniform appearance, but natural selection has favored different populations to mimic different models. So, what might drive the evolution of mimicry?

Conceptually, mimicry can be thought of as evolution involving natural selection by more than two species. Although it is convenient to mostly think of the ecology and evolution of species interacting in pairs, as in an arms race, most ecological communities have many species that interact in various ways. A phenomenon like mimicry forces us to come to grips with evolution occurring in this context of multiple species. Consider monarchs for the moment,

a

b

FIGURE 6.7. (a) A newly emerged monarch (*D. plexippus*) and (b) viceroy (*Limenitis archippus*), the toxic model with the less toxic mimic. (c) and (d) Their caterpillars do not look alike. This individual viceroy is the caterpillar I stumbled on in the field on May 14, 2001, and that emerged in my living room two weeks later.

involved in an arms race with milkweed and having evolved aposematic coloration to advertise their toxicity associated with sequestered cardenolides. Enter the viceroy butterfly, *Limenitis archippus*, also a Nymphalid (or brush-footed) butterfly (fig. 6.7), but whose close relatives in the genus *Limenitis* look very different from monarchs. If it were the case that the viceroy evolved by natural selection to look like monarchs, this would be a case of mimicry. Here is an evolutionary interaction between milkweeds, monarchs, predators of monarchs, and now an additional species that plays the game of looking a par-

c

d

FIGURE 6.7. (*continued*)

ticular part (like a monarch) to gain a particular benefit (reduced predation pressure from birds).

The first time I reared a viceroy butterfly, I had no idea what I was doing. As a new professor at the University of Toronto, I spent as much time outside of the city as I could get away with. During the summers, I lived at the university's field station, the Koffler Scientific Reserve, where I advised students and conducted my own experiments on monarchs and milkweeds, and where my partner, Jennifer, and I spent our free time wandering through the fields and forests. I would arrive at the field station on May 1 and live there for four months. One spring, I came upon a caterpillar on an aspen tree that looked like bird poop. It had all the slimy and mottled green and white goo you expect from a baby goose's behind. And yet I naively brought it into my living room and watched it grow to pupation, not knowing what species it was. Some two weeks later when it emerged from its chrysalis while we were having lunch, I certainly knew that it was not a monarch butterfly, and I instantly recognized that it must be a viceroy. I had finally experienced mimicry firsthand (see fig. 6.7b).

Going back to the early days of chemical ecology in the late nineteenth century, the same lepidopterists who were theorizing about sequestration of plant compounds into butterflies (Slater, Wallace, and Haase) were developing ideas of mimicry. In fact, the monarch-viceroy interaction was common lore of the time, as evidenced in the newspaper clipping shown in figure 2.4. It was the British naturalist Henry Walter Bates who crystalized the concept of evolutionary mimicry in the 1860s. Now termed Batesian mimicry in his honor, he envisioned the mimicking species as a sheep in wolf's clothing. In other words, the Batesian mimic takes advantage of looking like the toxic model species but does not invest in toxicity. As long as predators have learned to avoid the toxic model, the Batesian mimic will enjoy benefits without paying for sequestration or other forms of defense. The other common form of mimicry, termed Müllerian mimicry, and named after the German naturalist Fritz Müller, was conceived as an evolutionary interaction where both species are toxic, with natural selection reinforcing their resemblance, and both parties benefiting from predators having experience with the other.

In an important study, Jane van Zandt Brower, former wife of Lincoln Brower, who was also earning her PhD studying at Yale in the 1950s, demonstrated experimentally that viceroy butterflies were indeed mimics of monarchs. Viceroys on their own were eaten by naive birds, but not by birds that had an experience with monarchs. Van Zandt Brower also showed that viceroys were more palatable than monarchs, but clearly not as palatable as other butterflies, leading her to not categorize them as either Batesian or Müllerian mimics in the strict sense. This was a classic study and surely was the inspiration for Lincoln Brower's later obsession with monarchs, cardenolides, and mimicry. In further support of viceroys having evolved to mimic milkweed butterflies, it appears that the single species of viceroy (*L. archippus*) resembles three distinct milkweed butterflies in the genus *Danaus* in different parts of North America. This is highly convincing evidence for the evolution of mimicry, as it is exceedingly unlikely that other factors would favor the viceroy's having different coloration in different parts of its range, coincident with the coloration of three different *Danaus* species.

Viceroys are remarkable mimics of monarch models. In fact, several blunders have resulted from confusion between the two. Perhaps the most memorable error occurred in a 50-peso banknote issued by the government of Mexico in 2004. On this currency, which highlighted the region of the monarch overwintering grounds, several butterflies were depicted flying together, including a viceroy that was accidentally included among monarchs. The banknote was corrected and replaced in 2012.

SEQUESTRATION AND AUTOMIMICRY

There is one final piece to the story of mimicry in the milkweed butterflies, and it is what Lincoln Brower termed "automimicry," or mimicry within a species. Depending on what plant species a monarch eats, it will encounter tremendous variation in the abundance and polarity (water solubility) of the cardenolides it ingests. For a milkweed with trivially low cardenolides, an adult monarch will only have trace levels of cardenolides. As cardenolide levels rise in milkweed leaves, so too do they rise in monarch bodies, at least to a point. Beyond some intermediate level of cardenolides, about 3 milligrams per gram of dry mass, monarchs do not sequester higher levels of cardenolides. Sequestration saturates. This points to a likely burden of higher cardenolides, even to a highly resistant monarch. And what of the variation in chemical polarity? There is only so much that a monarch can do to metabolize, excrete, or otherwise selectively sequester particular cardenolides. Thus, both the amount and type of cardenolides in particular milkweed leaves will shape the specific compounds that are sequestered in the monarch's body.

Consequently, there can be tremendous variation in both the concentration and specific attributes of the cardenolides stored in monarchs. When Brower realized this, he argued that even within a single species, *D. plexippus*, there could be "mimicry." In this case, when feeding on a low-cardenolide milkweed, monarchs will only minimally sequester, will be highly palatable to predators, but may still benefit from looking toxic (the "automimic" or "self-mimic"; see the epigraph to this chapter). This automimic lacks cardenolides, and so acts as a Batesian mimic, but is the same species as toxic monarchs. Other individuals ("automodels"), having fed on higher-cardenolide milkweeds, reinforce the association of aposematism and toxicity. In either case, variation could be afforded within the species, as long as toxic models teach bird predators to avoid the orange, black, and white butterflies.

The combination of visual aposematism, the taste of cardenolides from butterfly wing scales, and consumption leading to vomiting can all reinforce the recognition of warning coloration. Once the caterpillar is done consuming milkweed leaves, cardenolides are concentrated in the gut, moved into the blood, and stored in the caterpillar's outer integument (or exoskeleton). Cardenolides eventually move through the chrysalis and appear in the adult butterfly. Where in the adult monarch those cardenolides end up is quite telling. They are highly concentrated in the wings and most concentrated in the wing scales. This pattern is concordant with two of the major rules of sequestration: put the toxic substances as far away from the internal organs as possible and, at the same time, make them as available as possible to a predator's taste buds, without the risk of loss of life. As a bird grabs part of a wing, inhales or tastes wing scales, and immediately rejects the bitter butterfly, the adult monarch has a reasonable chance of surviving. And the bird will remember the experience.

IT'S NOT ALL ABOUT THE ARMS RACE

When you are obsessed with coevolution, as I am, it is sometimes too easy to wear blinders and to interpret all the traits of an organism in this context. In fact, it is a general problem of evolutionary biology. Well, "problem," is perhaps too strong a word. This tendency is sometimes derogatorily referred to as "adaptationist"—interpreting all traits of an organism in terms of how we think it evolved. Some biologists are accused of being raving adaptationists, willy-nilly storytelling about why particular traits evolved. And it is all too easy to do so when the puzzle fits together neatly. I understand this criticism; we are scientists, after all, and depend on the rigorous testing of hypotheses and data. Nonetheless, I am a self-confessed adaptationist, at least happy to let myself imagine that evolution by natural selection is perfectly predictable through the lens of human eyes. But deep down, I know that I can only use this adaptationist lens to generate hypotheses, ideas that can then be tested with data and the scientific method.

For biologists studying arthropod-plant interactions, the traits of herbivores that seem to be involved in coping with plant defenses have rightly been interpreted in the context of adaptations involved in arms race coevolution. Could there ever be a nonadaptive explanation for caterpillars shaving trichomes and trenching latex canals? But even in such "obvious" cases, we need to exercise caution. As a case in point, consider the molecular changes in the monarch sodium pump that were introduced in chapter 3, making monarchs relatively insensitive to cardenolides. This is seemingly yet another easily interpretable adaptation. Presumably, monarchs evolved the sodium pump changes as an answer to cardenolides, allowing them to feed on such a toxic diet.

But starting in the 1970s, and continuing with casual observations over the past three decades, there has been some inkling among milkweed butterfly biologists that the sodium pump adaptations may have been more strongly driven by the evolution of cardenolide sequestration than to cope with the dietary cardenolides in the monarch's food. How to decipher such a question? Was food or predator avoidance the driver of the monarch's insensitivity to cardenolides? The evolutionary events of interest were all in the past, deep in the evolutionary history of milkweed butterflies. Nonetheless, such historical mysteries can be addressed. To answer this question, I was lucky to work with a talented and creative chemical ecologist, Georg Petschenka, who brought an uncommon diversity of skills to my lab, spanning evolutionary biology, physiology, chemistry, and deep roots in entomology.

When Georg arrived at my laboratory from Germany, he had just published a study showing that among milkweed butterflies, there were three progressive evolutionary stages of insensitivity to cardenolides. At the base of the milkweed butterfly phylogeny (see fig. 1.8), the earliest diverging milkweed butterflies maintained a highly sensitive sodium pump, similar to that of the rest of the animal kingdom. However, over evolutionary time, a group of milkweed butterflies (including the queen butterfly, *Danaus gilippus*) emerged with intermediate insensitivity to cardenolides. It would take millions of years, and the evolution of many additional milkweed butterfly species in that lineage, before yet another molecular change would occur, making the sodium pump even less sensitive to cardenolides than the ancestor or the queen butterfly. This final group contains relatively few species, including the monarch, *D. plexippus*.

What Georg knew, as did a few other naturalists, is that even the earliest diverging milkweed butterflies, those at the base of the phylogenetic tree, were quite capable of eating toxic milkweeds with high levels of cardenolides.

Then why did queens and, later monarchs, progressively evolve greater insensitivity to cardenolides? This nagging question was answered by taking a historical approach and by comparing the biology and chemical processing of a milkweed butterfly that is distantly related to the rest of the group (the "crow," *Euploea core*) with that of queens and monarchs (fig. 6.8; see also fig. 1.8). By comparing species that branched off at different times in the evolutionary history of the milkweed butterfly lineage, we gain a view into the changes that accompanied those branching events. Georg's results could not have been clearer. As the queens and monarchs progressively evolved greater cardenolide insensitivity, so too did they gain the ability to sequester cardenolides into their bodies. Consequently, cardenolide insensitivity was a means to reduce predation risk, not to tolerate cardenolides in their diet.

But, let's take a step back. The crow and queen butterflies, despite their more sensitive sodium pumps than monarchs, do not suffer from feeding on high cardenolide milkweeds. How do they do it? We don't really know, but we suspect that they have impermeable guts and don't take cardenolides into the body; they metabolize cardenolides that diffuse in, and they excrete other forms of the toxin. If the early diverging milkweed butterflies are "resistant" to cardenolides, then why the evolutionary sequence of more and more insensitive sodium pumps?

Coping with dietary cardenolides did not drive the evolution of the monarch's highly insensitive sodium pump. Instead, natural selection favored unprecedented levels of sequestration. To achieve such high levels of sequestration, cardenolides must be brought through the gut into the bloodstream of the monarch and stored in the exoskeleton (the hard outer skin). All of this requires exposing internal organs to the toxin. The natural selection imposed by bird predators, and probably other vertebrates as well, was so strong that it changed the course of how monarchs dealt with their toxic food. The ancestors of monarchs were already able to sufficiently cope with milkweed cardenolides in their food, but in later-evolving species, natural selection favored not just "coping" with them, but developing greater insensitivity, and hence being able to put

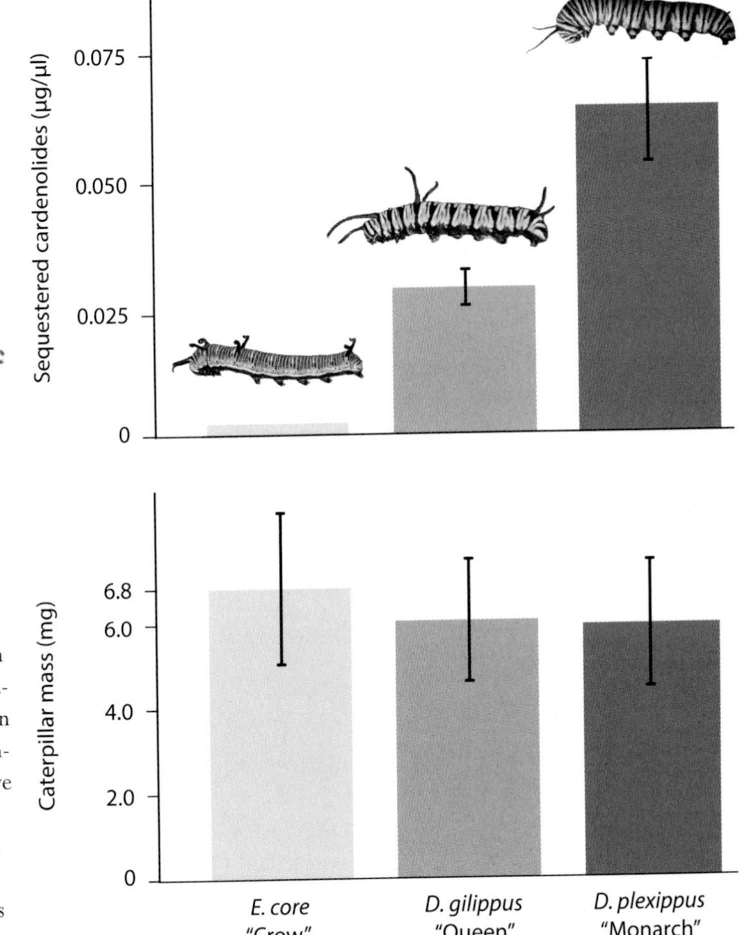

FIGURE 6.8. Among the milkweed butterflies, sequestration is highest in the monarch (measured as concentration in the blood) and intermediate in queen butterflies. The basal group, including *Euploea core*, which is sensitive to cardenolides, does not sequester them. Nonetheless, the growth rates of the three butterflies, even when feeding on highly toxic milkweeds, is more or less equivalent.

them to work in defense of monarchs themselves. It is currently unknown whether milkweed butterflies that do not sequester cardenolides have higher rates of predator attack or if they have found alternative defensive strategies. Addressing these issues would certainly make an excellent doctoral thesis.

Georg's result provided a profound insight. It showed that we cannot simply interpret traits that seem to fit the pattern of a coevolutionary arms race in that framework. The monarchs' sodium pump adaptations are not too far off, but the details are important. It was a third group of species, predators of monarchs, that drove their sodium pump to be resistant to cardenolides. Their highly insensitive pump allows caterpillars to bring cardenolides into their blood and body and to use them as a sequestered defense. This is how monarchs have managed to break that fundamental trade-off between growth and defense. And it is also how and why natural selection has engaged additional species, not only milkweeds and monarchs, but also bird predators and mimicking butterflies, in the evolutionary process.

For the growth of a caterpillar, perhaps the growth of any organism, to eat and avoid being eaten are two absolutely critical strategies. And for most animal species, engaging in one of these activities exposes the individual to limitations of the other. Feeding is dangerous and often exposes the animal to predators. Blending in and hiding has its merits, but it leaves the individual hungry. Monarchs do not appear to be subject to the typical conundrum of eat and be eaten (or hide and starve). They have intertwined their dining and defense. Monarch caterpillars do not hide; rather, they advertise their presence, feed during the day, and their poisonous food helps to protect them from bird predators. Being involved in a coevolutionary arms race allowed for this greater escalation and permitted predators to impose natural selection for sequestration because the caterpillars were already coping well with cardenolides. And such a relationship has had rippling effects in the community. It is not just queens and monarchs that are aposematically colored, but unrelated butterfly species have converged on the same coloration, mimicking the highly toxic monarch. Evolution, then, seems to occur at the level of an entire ecological community, starting with a two-species arms race, but extending its impacts onto other species.

How far and wide is the convergence and evolutionary impact of the monarch-milkweed arms race? This is the subject of the next chapter. For the moment, we will relax our focus on the monarch butterfly and peer into the rest of the milkweed's insect community. There lies a diverse set of species that all make their living feeding on our beloved bitter, toxic, and milky weed.

CHAPTER 7

The Milkweed Village

When we try to pick out anything by itself, we find it hitched to everything else in the Universe.

—John Muir, *My First Summer in the Sierra*

Summer in a milkweed patch is a colorful place. Not only are the flowers beautiful, but fragrances waft by, and bees are buzzing around. And you might see a monarch butterfly perched on a flower or find one of its caterpillars grazing leaves—as we know, they are not shy. And as you watch patiently, or perhaps as you turn over leaves, you are likely to be rewarded. Several species of other insects, with different shapes and sizes, but all with a familiar bright coloration also make milkweed their home. Two species of seed-eating bugs share many characteristics with monarchs. Not only are they aposematically red and black, but they also sequester cardenolides from the plant and have the same genetic substitutions in their sodium pumps. Turn to another plant, and bright yellow aphids may be sucking the milkweed's sap, or one of several beetles that are red and black may be climbing the stem. If the insects are chewing the leaves, you are likely to see them cutting the latex-delivering veins before they feed. It's a village or "community" of insects that are hitched together via a common resource. Their bed and breakfast is milkweed.

The coevolutionary arms race between monarchs and milkweeds is thus an exemplar of a widely repeated phenomenon. Looking into the milkweed community reveals a parallel set of species that have convergently evolved coloration, behavior, and sequestration. They participate in the arms race as a community. And as discussed earlier in this book, convergent evolution is a hallmark of adaptation, because multiple species have independently evolved to look or behave in similar ways (in this case specializing on milkweed). In this chapter, I will take a community ecologist's perspective on insect-milkweed coevolution, exposing how what begins with monarch butterflies extends to shape how multiple species interact. Some of milkweed's defenses are general and work to keep most of the community under control. Other defenses are highly specific and target particular insect species.

Summer in a milkweed patch is full of insects eating all the different parts of the plant, and in addition to sharing a history of battle with milkweed, they themselves interact in various ways. Drawing on John Muir's observation above, we will ultimately consider how the species that share milkweed as their food are interconnected.

THE CAST OF CHARACTERS

There are eleven species of insects that frequently attack common milkweed, and they all specialize in eating this plant, but to varying degrees. One species, the aphid *Aphis nerii*, will eat nearly any of the five thousand or so plant species in the botanical family Apocynaceae (it is nonetheless restricted to this one family), while another aphid species, *Myzocallis asclepiadis*, appears to be restricted to only the common milkweed (*Asclepias syriaca*). The milkweed herbivores are quite taxonomically diverse, and as we will see, they include a fly (order Diptera), two Lepidopterans (the monarch butterfly and a moth), a few beetles (order Coleoptera), and several true bugs (order Hemiptera). Together, this means that some 350 million years since the major taxonomic groups of insects separated, there have been individual species or lineages that have independently colonized and adapted to eating and living on milkweeds.

Some 350 million years ago, at the border of the geological epochs known as the Devonian and the Carboniferous, there were two massive landmasses on the planet: Gondwana and Laurasia. Flowering plants had not yet evolved (plants consisted of algae, mosses, and tree ferns), and although insects were abundant, they were just beginning to evolutionarily diversify. The insects mostly consisted of wingless forms like silverfish; the first winged species, such as dragonflies, were just beginning to emerge. The first fossil evidence of insect herbivory is linked to this period (300–400 million years ago), but those insects were rather unlike the beetles, bugs, and caterpillars of today.

Around this time, insects experienced a tremendous evolutionary development, multiplying and radiating from few into millions of species. Just as Darwin had envisioned, from a single ancestor, lineages branched off and were henceforth evolutionarily independent of other lineages that subsequently branched off. About 350 million years ago, one branch split off that would evolve into the true bugs, which lack metamorphosis (Hemiptera, sucking insects including aphids and Lygaeids), and another branch split off that would be the only insect group to evolve complete metamorphosis (a megagroup called the Holometabola). From the Holometabola, many lineages subsequently branched off: the beetles (Coleoptera) around 300 million years ago, the flies (Diptera) around 175 million years ago, and the butterflies and moths (Lepidoptera) around 150 million years ago. In each of these very different groups, each with distinct ecologies and feeding behaviors, a choice few insects evolved that utilize milkweeds as food. Even within a group—for example, the beetles—multiple species colonized the milkweeds. It is almost miraculous that on so many of the branches of the insects' diverse evolutionary tree, each consisting of thousands of species, lies a species specialized to eat milkweed.

But this insect community, any way you slice it, is a bit depauperate. In other words, for such a common and abundant plant, eleven species of insect pests are very few. Take for example, a plant species that frequently co-occurs with milkweed in eastern North America, the tall goldenrod, *Solidago altissima*. One of my predecessors studying plant and insect ecology at Cornell University, Richard "Dick" Root, who studied both goldenrod and milkweed, reported more than one hundred species of insects that feed on tall goldenrod (excluding flower visitors). Dick was himself tall and gregarious, and a disciplined naturalist always looking for an excuse to go out and observe insects on plants.

Based on literature and his extensive field research, he noted that forty-two of the one hundred species found on goldenrod are specialist insects that feed only on goldenrod or closely related plants. This stands in contrast to milkweed's depauperate insect fauna (a total of eleven species, all specialists). It seems that the high level of toxicity and the gluey latex barrier, driven by strong coevolutionary interactions with monarchs and other milkweed insects, restricts the numbers in this insect community.

When characterizing an ecological community, there are many ways to describe it, with the number of species (termed species "richness") being the simplest metric. Members of a community can also be characterized by the extent to which they are picky eaters. As we know, monarchs are termed "specialists" because their diet is restricted to milkweeds. In fact, all of the milkweed herbivores are specialists. As mentioned above, evolutionary relatedness is still another way to categorize a community. The eleven milkweed insects belong to eight distinct taxonomic families (the three aphids, for example, are all in the same family, Aphididae). Compare this with the forty-two specialist insects found on tall goldenrod, which encompass seventeen distinct taxonomic families, let alone the more than sixty generalist insect species. The final means that I will mention for characterizing a community is what is called "guild structure"—a way to group species by the way they interact with their environment.

In the same year (1964) that Dick Root earned his PhD from the University of California–Berkeley, studying bird foraging and guild structure, he started his career as a plant-insect ecologist at Cornell. As part of his doctoral thesis, Root was the first to define the ecological guild, as "groups of species that exploit the same class of resources in a similar way." Root left this definition purposely flexible, as that "class of resource" could refer to milkweed versus goldenrod, tree foliage versus tree bark, or grasses versus mosses. And "in a similar way" could be equally squishy. Root was a philosophical man and a firm believer in the need for ecologists to grapple with the complexity of nature. He wrote: "Part of the art of becoming an ecologist involves developing a set of attitudes for coping with the complications that stem from the individualis-

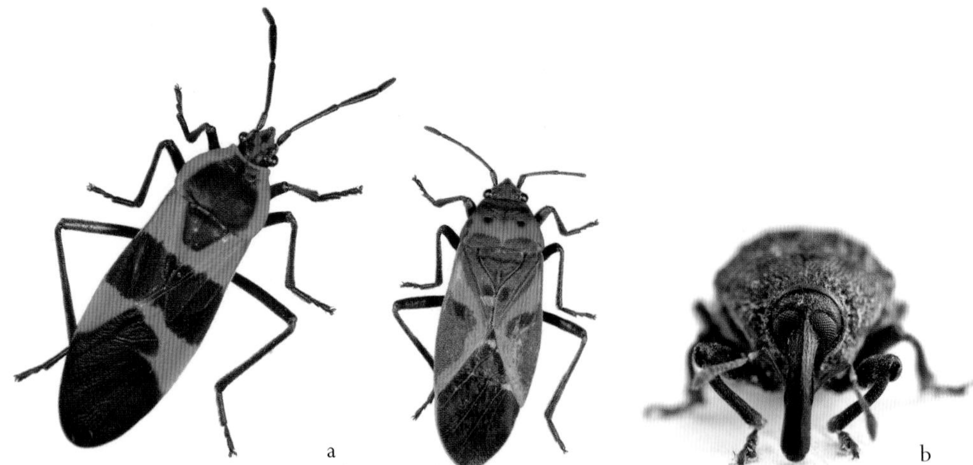

FIGURE 7.1. Three seed-eating milkweed insects: (a) The large and small milkweed bugs, *Oncopeltus fasciatus* and *Lygaeus kalmii*, and (b) the weevil, *Rhyssomatus lineaticollis*. The feeding beaks of the bugs are not visible because they are folded down along their undersides. The black coloration of *Rhyssomatus* is probably not aposematic, as weevils readily drop from plants, and they are very difficult to find on the ground, where they blend in with soil and tiny rocks.

tic nature of species. One of the most important of these is an ability to match the question one is asking with the most appropriate grouping of species—the set that will reveal valid patterns that act in nature."

For the milkweed insects, the general resource is clear, but how will we define the ways in which the resource is exploited? Like all plants, milkweed is not a single food: there are the roots, shoots, and fruits. Let's start by dividing up the plant, top to bottom, into its main distinct parts: seeds, fruits, flowers, leaves, stem, sap, latex, and roots. The milkweed insects utilize all of these parts, although none exclusively eat the flowers or latex.

Seed Eaters

Starting at the top of the milkweed, where the seeds develop, we find three species of insect herbivores: two true bugs, the large and small milkweed bugs (*Oncopeltus fasciatus* and *Lygaeus kalmii*, respectively) both in the family Lygaeidae, and a beetle, *Rhyssomatus lineaticollis*, in the weevil family (Curculionidae) (fig. 7.1). These insects have a lot in common in addition to their black coloration and penchant for milkweed seeds. They have mouths at the end of "beaks" that are several millimeters long, and although they all have wings, they very rarely fly.

Adult weevils are active both in the spring and in early autumn. In the spring, however, their larvae are consumers of the stem tissue (before milkweeds have seedpods), while in the late summer their larvae are inside the pods, feasting on maturing seeds. As for the two bugs, they are indeed seed feeders, with nymphs and adults eating mature seeds when pods open. Presumably, in an evolutionary response to these feeders, milkweed seeds have triple the cardenolide concentrations of leaves. The seed-feeding guild thus has a few species that are similar, but there are important differences in when and how they attack seeds. The plant likely has specific defensive adaptations to these seed feeders that are yet to be discovered.

Suckers

Aphids are commonly (and not so affectionately) known as plant lice. They suck the sap of plants, making the plants sticky and sometimes wilty. Tiny and yet voraciously hungry and prolific, aphids are pests of farms and wildlands alike. And milkweed is home to three species: *Aphis nerii*, *Aphis asclepiadis*, and *Myzocallis asclepiadis* (fig. 7.2). These small suckers can be very abundant. But perhaps because they are small, and two of them are greenish, like the plants they attack, when I started working on milkweeds in the year 2000, there was not a single published study on the ecology of either *Aphis asclepiadis* or *Myzocallis asclepiadis*. Consequently, my lab group dove in and has been studying the similarities and differences among the aphid guild ever since.

When several species are in the same guild, it is thought that each species must have important distinctions that allow them all to coexist. If there were very few differences, we might expect them to compete very strongly and for the plant to perceive all aphids as one group of suckers in the coevolutionary arms race. And although this group of aphids exploits the same plant part (phloem sap) in a similar way, the three species exhibit diverse strategies: *Aphis nerii* and *Aphis asclepiadis* feed in tightly packed clusters, and preferentially near the top of the plant. In contrast, *Myzocallis* feeds in a dispersed pattern, with some personal space between each aphid, on the undersides of the lower leaves. Despite the fact that aphids avoid puncturing the latex-delivering canals and they all have to cope with cardenolides (which flow in the phloem sap), their differences suggest that milkweed may have to use different strategies to defend against the three aphids.

Among the aphids, *Aphis asclepiadis* is nearly always tended by ants (see fig. 7.2b). Ant-tending of aphids is one of those special ecological interactions, a "mutualism," where both species benefit. Ants drink the sugar-rich excrement of aphids (not called "frass" in this case, but rather "honeydew"), and in turn protect the aphids from predators. Interestingly, among this guild of milkweed aphids, ants pay the most attention to *Aphis asclepiadis*, and studies in my laboratory showed that the presence of ants allows their populations to grow larger than if ants are absent. Milkweed's defense against *A. asclepiadis* involves traits that reduce ant tending. For example, Tobias Züst, a former postdoctoral fellow in my laboratory, recently discovered that feeding on higher cardenolide plants resulted in higher cardenolide honeydew and reduced ant-tending of *A. asclepiadis*. Regarding the other two aphids, *A. nerii* is typically not tended by ants, and *Myzocallis* often has ants feeding on its honeydew, but on the leaves below where the aphids are feeding. Thus, ants do not "tend" *Myzocallis*, and we have shown that ant-feeding on their honeydew does not benefit the aphid.

Our studies of the aphids have also revealed other important ecological differences between the species. Although each species is specialized enough to collect and sequester cardenolides in its body, only *Aphis nerii* is aposematically colored (bright yellow), while *Myzocallis* and *Aphis asclepiadis* are relatively drab and inconspicuous. I considered *Myzocallis* a "fugitive" species, whose specialty is movement, while the other two species have distinct attributes that are their

a

b

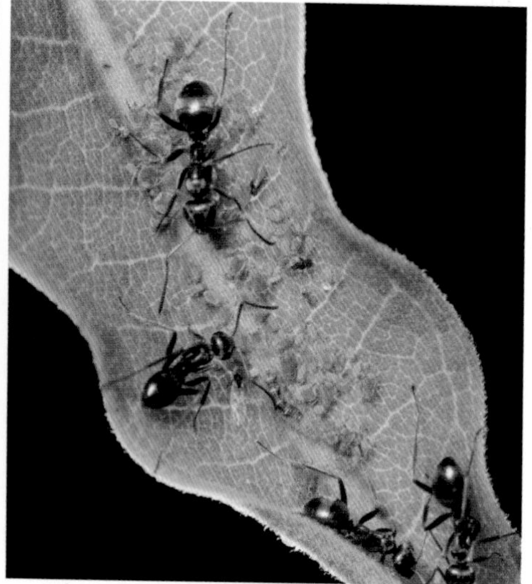

c

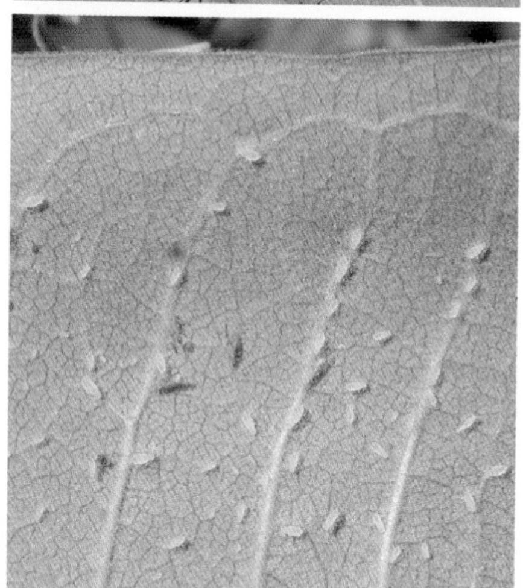

FIGURE 7.2. The three milkweed aphids: (a) *Aphis nerii* (aposematically colored bright yellow), (b) *Aphis asclepiadis* (tended by ants), and (c) *Myzocallis asclepiadis* (which feed in a dispersed fashion on the undersides of leaves; note adult winged individuals). *Myzocallis* is the most variably colored aphid and can be pink to yellow to clear and sometimes appears spotted. A winged adult of *Aphis asclepiadis* is near the middle ant, and a predaceous syrphid fly larva is next to the winged aphid. These three aphids are also abundant at different times of the summer (see fig. 7.7).

specialties. Indeed, the different strategies among the milkweed aphids, one each relying on ant defenders, high mobility, or aposematic coloration, likely allows the three aphids, all in the same guild, to coexist. From the milkweed plant's perspective, one aphid is not any other aphid, and the arms races necessitate some specificity in how milkweeds defend and how the aphids exploit.

Chewers, Miners, and Borers

The rest of the milkweed insects are "chewers," although they chew plant parts in slightly different ways. Two species are especially worthy of note. First is one of the least-studied insects on milkweed, a fly that feeds between the layers of a leaf, physically concealed from the external world, but leaving behind an evidence trail of its feeding activity (fig. 7.3). This "leaf miner," *Liriomyza asclepiadis* (Agromyzidae), can be very abundant (tens of miners per plant) and produces visible splotches where individuals have been feeding (up to 4 square centimeters on leaves). We know virtually nothing about how this miner has engaged in the arms race, nor what milkweed has done in defense. Leaf mining is an especially interesting guild, with only one known species on milkweed, chewing the leaves between the upper and lower epidermal layers, with its flattened body. More generally, leaf mining is a strategy that has evolved numerous times among insects that feed on plants other than milkweed, including in flies, beetles, moths, and wasps. Because leaf miners feed between the leaf layers, they are less apparent to their enemies in some ways, but they are also trapped and cannot run away or drop off the plant. Parasitoid wasps often inject the larval miner with their eggs through the leaf epidermis, and predators will occasionally rip open the leaf to devour the tiny morsel. If the larval fly avoids these insults, it will emerge from the mine only at the end of its larval stage, to drop to the soil and pupate. Because of the specific mode and location of the miner's feeding, I suspect that some general plant defensive traits (like cardenolides) are effective against the miner, while other defenses may be specific to their more private arms race.

The four-eyed red milkweed beetle (*Tetraopes tetrophthalmus*) is in the family of long-horned beetles, Cerambycidae. "Four-eyed" is no joke; they are called this because their antennae completely bisect their eyes, and as a consequence they have four functioning eyes (fig. 7.4d). The "why" behind these four eyes is unclear, and simple experiments that cover individual eyes and study beetle behavior would certainly be fruitful. These beetles can be very dense in the

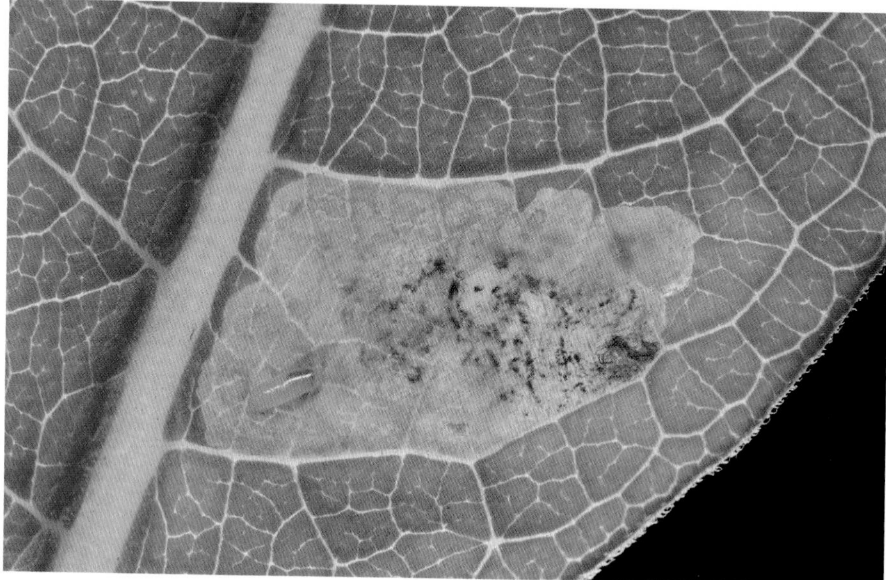

FIGURE 7.3. The larva of the milkweed leaf mining fly, *Liriomyza asclepiadis*, here removed from its mine. Dark spots in its body are chewed leaves in its gut, while the spots on the leaf are its excrement or "frass." Although this is the only known leaf miner on milkweed, there are likely other closely related but undescribed species that have evolved to attack other milkweed species.

field, with upward of twenty adults per square meter of milkweed plants, and like monarchs, they are easy to spot during the day, as they feed in the open and show off their red and black coloration.

Adults chew on leaves and flowers, while larvae burrow down in the soil and feed on milkweed roots. Although *Tetraopes tetrophthalmus* primarily feeds on common milkweed, *A. syriaca*, there are at least twenty-four other species of *Tetraopes*, most of which have specialized on other milkweeds. *Tetraopes* is a group that does not easily fall into a single guild, since it has its greatest impact as larvae feeding on roots, but adults feed on the aboveground plant parts. For adults, trichomes and latex appear to be the strongest plant barriers—a good investment for the plant, since these traits also thwart most of the chewing insects. But root defense is another matter entirely, because milkweed roots have neither trichomes nor latex. Milkweed roots do have cardenolides, which negatively impact *Tetraopes*, as well as volatile organic compounds that are released after attack and that attract predatory nematodes. Thus, the community of insects on milkweed has imposed natural selection for some generally effective plant defenses, yet each species' specific traits have also placed unique evolutionary selection on the plant.

EVOLUTIONARY HISTORY AND CONVERGENCE

There are two other milkweed insects that I have not yet introduced. They too are strictly leaf chewers, like the monarch. As such, I chose to introduce them here because, although they are unremarkable in the guild to which they belong, they vividly exemplify aspects of convergent evolution that are important for our understanding the insect community's joint arms race with milkweed. The misnamed milkweed tussock moth, *Euchaetes egle* (Arctiidae, misnamed because "tussock moth" is the common name of a different family of moths), and the milkweed leaf beetle, *Labidomera clivicollis* (Chrysomelidae), share coloration and behavior with the rest of the milkweed fauna despite being distant relatives (figs. 7.4a, 7.4c). The highly contrasting black against bright colors shows up over and over again.

At the beginning of this chapter, I mentioned that the milkweed fauna assembled over the eons from lineages that diverged from a common ancestor some 350 million years ago. Understanding this point is essential to understanding the nearly unimaginable nature of convergent evolution in these insects. Let's take a little time to pick this apart. The milkweed weevil (see fig. 7.1b), four-eyed beetle (figs. 7.4b, 7.4d), and leaf beetle (fig. 7.4c), are each beetles from different taxonomic families, and each colonized milkweed and evolved its coloration, behavior, and physiological adaptations independently. In other words, imagine the ancestor of the beetles branching off from the rest of the insects some 300 million years ago. That lineage kept splitting, there are now some five hundred families of beetles alone, and each one of these families

FIGURE 7.4. Three aposematically colored specialist leaf-chewing insects, all feeding after having deactivated the latex defense system: (a) The milkweed tussock moth, *Euchaetes egle*, has notched the leaf midrib such that it is hanging down. (b) The female *Tetraopes tetrophthalmus* (flat against the leaf) made several cuts to the leaf midrib before the male arrived for mating (note latex flowing out at the lowest cut). (c) The milkweed leaf beetle, *Labidomera clivicollis*, has clipped several smaller side veins (not visible) before feeding. (d) The four eyes of *Tetraopes*, each pair bisected by an antenna.

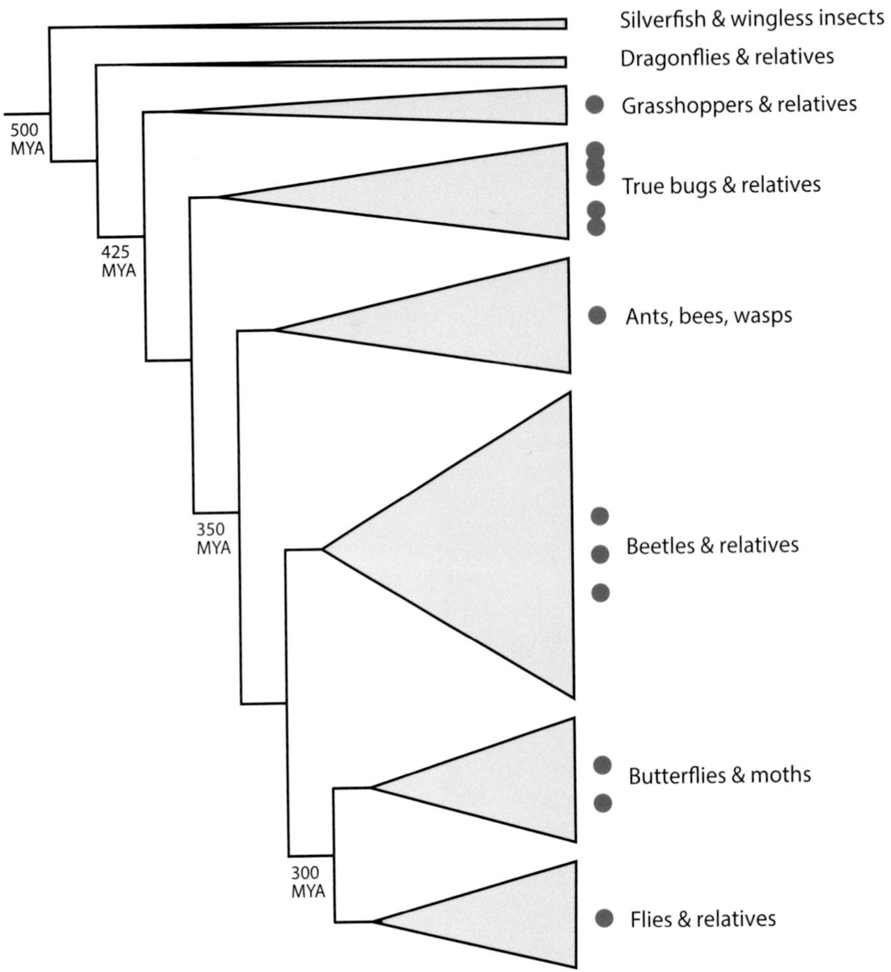

FIGURE 7.5. A representation of the phylogeny (evolutionary history and relationships) of the insects. Note that the insects originated about 500 million years ago (MYA). The size of each triangle indicates the number of currently recognized species in that grouping (e.g., about 100,000 true bugs and 350,000 species of beetles). Green dots indicate the independent evolution of specialized feeding on common milkweed (*Asclepias syriaca*). Note, for example, that five different true bugs (three aphids and two lygaeids) have colonized *A. syriaca*. Red dots indicate the independent evolution of insects from distinct groups that have colonized cardenolide-containing plants (grasshoppers on an African milkweed and a sawfly wasp on *Helleborus* plants in the family Ranunculaceae). Overall, hundreds of insect species, many of which are currently unknown, have colonized and specialized on cardenolide-containing plants. The independent colonization of milkweed and evolution of these insects in six groups (taxonomic orders) represents one of the best examples of convergent evolution.

became its own branch and split into a multitude of genera and species. And if we estimate one million species of beetles in total, then on average, each of the five hundred beetle families has two thousand splitting branches—each (fig. 7.5)! The three milkweed beetles are mere blips among the million, in different familial lineages, and have independently arrived at their traits, as the bulk of their close relatives have no association with milkweed at all.

In contrast, when we zoom in on a group of closely related beetles, say the twenty-five four-eyed species in the genus *Tetraopes*, they share many traits: how they look, their behaviors, and their choice of food plants. But this commonality is not independent among the *Tetraopes* species; because they inherited the traits from a recent common ancestor, all the descendants share many traits. When single species or small groups of species among the sea of all insects arrive at the same host plant and evolve the same traits (while each of their close relatives are doing something different), we can infer that the various disparate species have experienced convergent evolution.

DÉJÀ VU

The common warning coloration of most of the milkweed insects is plain as day, and most sequester cardenolide toxins from the plants. Perhaps less obvious is that all species of milkweed insects have more or less converged on avoiding latex. I say more or less because the seed-feeding bugs simply do not encounter latex, as is the case for the aphids, which feed between cells and do not puncture the latex delivering canals. It is less clear how the leaf mining fly avoids latex, but its stealthy feeding between leaf layers must also avoid puncturing the latex canals. To me, most astounding is that the larvae of the butterfly, moth, weevil, long-horned beetle, and leaf beetle each independently evolved very similar trenching behavior. That four milkweed feeders use this behavior was described by David Dussourd, then a graduate student at Cornell, and his adviser, Thomas Eisner, as they wrote in 1987 that trenching "is performed as a matter of dietary routine" by several milkweed insects that chew the leaves and encounter latex. They went on to describe how these insects typically first inflict bites that drain latex and then feed on the leaf tissues beyond the cuts that are latex-free.

There are subtle differences in how the milkweed chewers deactivate milkweed's latex defense. As described in chapter 5, newly hatched monarchs cut their circle trenches and later in development notch leaf petioles or midribs. *Rhyssomatus* weevils chew through the middle of leaf mid-veins and then feed in a dispersed pattern beyond the area that is chewed. Some, like *Tetraopes*, with wide mandibles, repeatedly clip the central mid-vein of leaves, and still others, like *Labidomera* clip the smaller side-veins on the sides of leaves. *Tetraopes* adults usually leave very characteristic damage; they leave the tips of milkweed leaves in the shape of a "U" or "V" because they feed on the tissues that have been denuded of latex by cutting the mid-vein (see fig. 7.4b). This shape results from the fact that the far edge of leaves beyond the clipped veins still receives latex from canals that are curving up the leaf's edge. In addition, once a deactivated leaf is abandoned, other herbivores, like generalist slugs, which typically reject milkweed, move in and feed on the latex-free areas of the leaf (fig. 7.6). Both *Tetraopes* and *Labidomera* also usually make several cuts beyond the first one, seemingly to be sure that latex flow is actually diminished. Despite some diversity in how they do it, all the milkweed chewers deactivate latex behaviorally, they all cut veins with their mouthparts, and they all use a sequence of behaviors ending with feeding on leaf tissues beyond the location where the canals were cut.

To chew on latexy leaves requires deactivation of this plant defense, and a set of similar behaviors has indeed independently evolved several times in conjunction with feeding on milkweed. If latex was not an important selective agent on insect herbivores, we simply would not expect to see such repeated evolution of parallel behaviors. Each of the milkweed chewers started with a different set of traits in terms of size, type of mouthparts, and preferred feeding location, but to cope with the latex problem, the same general solution emerged. Similarly, to not be poisoned by milkweed's cardenolides, adapta-

FIGURE 7.6. A generalist slug, not a typical herbivore of milkweed, here feeding on common milkweed (*A. syriaca*), taking advantage of the latex canals' having been previously severed by a *Tetraopes* adult. Slugs and snails also occasionally feed on lower leaves when they are turning yellow and have highly reduced latex flow.

tions are needed to prevent them from binding to the sodium pump (outlined in chapter 3). It would seem that a syndrome of aposematic coloration, deactivation of latex, and tolerance of cardenolides all evolved convergently in the milkweed insects.

To push this pursuit of understanding convergent evolution further, in a decade-long collaboration led by Susanne Dobler at the University of Hamburg (Germany), we examined the genetic sequences of the sodium pump in eight of the eleven insect herbivores of common milkweed, several other milkweed insects from across North America, as well as a comparison group of insects that do not feed on milkweed. Susanne is a careful and rigorous evolutionary geneticist who is driven by patterns she observes in the field. Our goal was to address the genetic basis of adaptations to cardenolides and to ask whether the specific genetic changes that allowed monarchs to tolerate and sequester cardenolides were the same genetic changes in other, unrelated insects. We were astonished by the results.

We discovered the same genetic mutations found in monarch butterfly sodium pumps have evolved in four additional species: *Oncopeltus fasciatus*, *Lygaeus kalmii*, *Labidomera clivicollis*, and *Liriomyza asclepiadis*. Thus, a butterfly, two true bugs, a leaf beetle, and a leaf mining fly all found the same solution—the same exact solution, right down to the fine-scale molecular changes in the DNA encoding for the sodium pump to crack the code of cardenolide defense. Convergence at this level seemed almost too perfect. Why would the exact same genetic substitution keep evolving?

The answer has to lie first in whether there is actually more than one potential solution to the problem of resisting cardenolide poisoning. We know there is more than one solution (indeed some milkweed herbivores use different molecular changes). Second, if there is more than one solution, natural selection will be guided by the cost-to-benefit ratio of the different means of cardenolide resistance. By this, I mean that natural selection will favor those solutions with the highest benefit after taking into account any costs. Take, for example, the idea that there are many genetic base pairs coding for the shape of the cardenolide binding pocket in the sodium pump. Two different genetic substitutions may be equally beneficial in terms of reducing cardenolide binding, but one of these substitutions may be much more costly, perhaps because it reduces the functioning of the sodium pump (see chapter 3 for an explanation of sodium pumps). Alternatively, a particular substitution may profoundly reduce cardenolide binding because it changes the shape of the binding pocket at precisely the right spot, compared with some other substitution that may not have such great benefits. Natural selection works to maximize benefits and minimize costs. And all we know at this point is that a small set of solutions to the cardenolide problem seem to have a very low cost-to-benefit ratio, thus repeatedly appearing in different milkweed insects. Some of the milkweed insects do have alternative genetic substitutions in their sodium pumps, and it is yet to be discovered why these have been favored over others.

To begin to identify the costs of a cardenolide-resistant sodium pump, my laboratory has teamed up with Noah Whiteman of the University of California–Berkeley, a bold and upbeat evolutionary geneticist and connoisseur of molecular adaptations of insects on plants. Noah is a rare breed among biologists, equipped with all the tools of modern molecular biology, with a keen sense of organisms and how they behave in nature, and always up for a challenging scientific problem. Together our goal is to genetically manipulate an insect species to have the different sodium pump forms, including the highly conserved sensitive form found in most animals, the few types of resistant forms from milkweed insects, as well as other forms predicted to be resistant to cardenolides but rarely found in nature. By using the fruit fly, *Drosophila melanogaster*, which is used in thousands of genetics laboratories across the world and therefore has very well-developed protocols, we hope to create living fly colonies that are genetically similar, but have specific changes in their sodium pumps. Then by rearing the flies under different conditions, with and without cardenolides, at low and high competition, and under lush or limiting food resources, we will investigate the costs and benefits of particular genetic substitutions in the sodium pump. We expect that such a highly manipulative approach, which would only be possible with fruit flies, will help explain the pattern of molecular convergence we see among the naturally occurring insects that feed on milkweed.

Several mysteries remain. Why do some milkweed insects not evolve sodium pump genetic substitutions, and what is their alternative strategy to coping with cardenolides? Yes, there has been convergence among insects in several distinct lineages, but there are also species like the Arctiid moth, *Euchaetes*, which neither sequesters cardenolides nor has any substitutions in its sodium pump (but it is still aposematically colored). This uncoupling of sequestration and aposematic coloration is ripe for further study. Other insects, like *Labidomera* leaf beetles and *Liriomyza* mining flies, have genetic substitutions in their sodium pumps but do not sequester cardenolides. Worldwide there are easily hundreds, probably thousands, of species of insects that eat cardenolide-containing plants. Much work is yet to be done to understand why some insect species convergently evolve the exact same molecular solutions to cardenolides, while other species evolve alternative and highly divergent strategies.

A MILKWEED BY ANY OTHER NAME

Our understanding of convergence and its importance for evolution are reinforced by the depth and breadth of the evolution of insect traits. By depth, I mean how similar are the convergent traits: do the species simply look the same, or did they evolve by the same molecular mechanisms? The latter was certainly the case with the sodium pump evolution of milkweed insects, which shows deep molecular convergence among several species. And by breadth, I mean the extent to which highly unrelated organisms have converged on the same solution. Again the milkweed insects across many evolutionary lineages have certainly converged. But convergence can be considered at many scales. For example, are there parallel universes, or at least villages, of insects on other milkweed species? For the many *Asclepias* species across North America, there is a shared and parallel insect community. Some species, such as monarchs, *A. nerii*, and *Lygaeus*, are relatively promiscuous and occur on many milkweed species, while others species, like *Tetraopes* and *Euchaetes*, although host-specific, have close relatives in the same genus that have branched out onto other milkweeds. But what happens when we go further afield—specifically, if we consider an Old World relative of milkweed called *Calotropis*, a latex-rich plant that grows natively in the Middle East and India? The insect community on *Calotropis* reveals an assemblage that has converged and parallels that of the common milkweed of North America.

Before we get there, however, I must explain a mystery that relates to the story of *Calotropis*. The story has to do with how common milkweed, a native of eastern North America, got its scientific name, *Asclepias syriaca*. The species name might suggest, confusingly, that the plant is native to Syria. To understand this appellation, we need to dig back to the 1630s, when a French botanist and physician, Jacques-Philippe Cornuti, was studying plants imported from North America to the Paris Botanical Garden. Cornuti examined the plant we now call common milkweed and concluded, in his 1635 book, *Canadensium Plantarum Historia*, that it was the same species then called "Beidelsar," a plant known to the Middle East that had been classified as *Apocynum syriacum* by Carolus Clusius in 1601. Unfortunately, Cornuti did not see the two plants' flowers, which are quite different. More than a hundred years later, in his 1753 opus *Species Plantarum*, Carolus Linnaeus kept the species name *syriaca* for common milkweed, although he put the plant in the new genus *Asclepias*, which he named after the Greek god of medicine. Linnaeus recognized that *A. syriaca* was a New World plant native to eastern North America, but he decided to keep the old species name, probably for consistency. Additionally, he named Beidelsar *Asclepias gigantea* (which would later be renamed in the genus *Calotropis*). However, Linnaeus could have straightened things out, had he realized that Cornuti had confused the two plant species as one. But, Linnaeus was unaware of Cornuti's mistake, and unaware that Clusius had already classified Beidelsar as *Apocynum syriacum*. In my humble opinion, Linnaeus should have kept the species name *syriaca* for Beidelsar, retaining some historical precedence and a geographically correct name, but he apparently did not check his sources closely enough.

Calotropis, as it is now known, has two species, one native to Syria and neighboring countries in the Middle East (including northern Africa) and the other native to India. Looking somewhat like common milkweed, it also behaves in a similar way. It is a weedy species that has colonized many places, including widespread invasion in Australia, Brazil, and the Caribbean Islands. And there are some other similarities as well. Up to four Asian milkweed butterflies (in the tribe Danaini, but in several genera) feed on *Calotropis*, as do several Lygaeid bugs (relatives of *Lygaeus* and *Oncopeltus*, but again in different genera). In addition, as on North American milkweeds, there are aposematically colored Arctiid moths on *Calotropis*. There are some four species of aphids on *Calotropis*, including *Aphis nerii,* which parallels the group of aphids on common milkweed. *Calotropis* also supports a seed-feeding weevil that is only a distant relative of its North American counterpart, but is similarly colored. Other specialist seed predators of *Calotropis* include two aposematically colored flies in the genus *Dacus*. It is quite a convergent village of insects.

Among the major insect groups, what has been conspicuously missing from the milkweed faunas we have discussed so far are grasshoppers (Orthoptera) and wasps (Hymenoptera), two major groups that include herbivorous insects. Grasshoppers have colonized *Calotropis*, and in both North Africa and Asia there are species in the genus *Poekilocerus* that have specialized on cardenolide-containing plants. Yes, these species are aposematically colored, sequester cardenolides, and have largely insensitive sodium pumps. As far as we know, herbivorous wasps have not colonized milkweed, but a sawfly wasp in Europe, *Monophadnus latus*, has colonized toxic *Helleborus* plants, which produce toxins very similar to cardenolides. And, yes, this wasp has converged on the same genetic sodium pump substitutions as the monarch and several other milkweed insects. During the past 350 million years of evolution, as insects diverged into six major groups that include plant feeders (see fig. 7.5), each group independently evolved species that would specialize on these toxic plants.

That is a lot of convergent evolution. Not only do similar forms with similar behaviors and physiologies evolve to utilize the same host plants, but in distant places, where similar plants have been evolving independently for a long time, there too have similar insect communities evolved. The extent to which there are similarities and distinctions in the respective arms races across continents is an open question, and one that would be fascinating to unravel. The final aspect of milkweed's insect community that I want to discuss is that of the entangled bank of species, and how they are ecologically interconnected, despite representing more than 350 million years of independent evolution.

HITCHED TO EVERYBODY ELSE IN THE VILLAGE

When insects share a host plant, they are bound to interact with each other, either directly or indirectly. This is likely true in any community that relies on a shared resource base. For example, the organisms that live on a bed of moss or the microbes that live in our gut must be in physical contact with each other and also, at least occasionally, compete with each other for food. As plants are a major food resource, they are a stage on which many such ecological interactions among herbivorous insects play out.

Interestingly, it is probably a relatively rare event for insects on a shared host plant to be limited by plant tissue itself. As is often quipped by ecologists, "the world is green," and one logical conclusion of this is that insects on plants are unlikely to directly compete, because there is usually enough green. Why the world is green is the subject of much debate among ecologists, but it likely has something to do with the facts that plants are toxic and most insects out there are limited by their predators, rather than being limited by plant tissues they wish to consume. Insects on milkweeds are no different.

So how do insect herbivores interact on a plant if not through competing for limited plant tissues? As we learned in chapter 5, plants are not static organisms, and once fed upon by one insect species, they can change in terms of their attractiveness and nutritive quality. In fact, there may be competitive interactions between the insects on a plant (where the presence of one species has a negative impact on another), even if 90 percent of the foliage and roots remain intact. This is because damage by one insect induces defenses that may then reduce the preference or performance of a subsequently feeding insect. It is such induced defenses that create dynamism and can structure the interactions between insects. Changing plant quality is one of the ways in which the milkweed community of insects is hitched together.

There are two other aspects of this connectedness of the milkweed insects that need explanation. First is the consistent climatic (seasonal) sequence, or what ecologists call phenology, in which certain insect species are most abundant. In other words, some of the insect species are likely to emerge and damage the plants early in spring, while others are more likely to emerge late in the summer. Phenology is relevant to understanding insect community structure, as those early-season species should disproportionally affect the later-season species (not vice versa), because those that arrive first and start feeding on clean plants induce changes in plant quality for later feeders. The second aspect has to do with whether there are specific plant responses to different feeders. If there is no specificity, then insect community structure could be understood based simply on when plants are first damaged. In contrast, if different species each change the plant in distinct ways, then we must understand these specific changes as well as whether the insect species are equally affected by the different changes. The latter scenario is generally the case for plant-insect interactions, and for milkweed herbivores in particular.

It is complex and messy, but that is apparently the way nature, or at least the ecology of plants and insects, is organized. Here is what we know. In central New York State, and in much of the northeast, there are essentially three seasonal waves of milkweed insects (fig. 7.7). Some of the seasonality is determined by the availability of certain plant parts. For example, the seed-feeding bugs are most abundant in autumn, when milkweed seedpods are opening. Other late-season insects like the bright yellow aphid, *Aphis nerii*, arrive late on the scene because they blow up north from southern climates where they spend the winter. *Aphis nerii* does not "migrate" because it does not make the return trip, but it cannot withstand winters, and so each summer the population expands north from southern regions. Like the monarch, it is essentially a tropical insect. Also like the monarch, as the season progresses, *A. nerii* moves north with the availability of milkweed. However, there is no indication that this northward movement is due to directional movement, and there is no return trip. In fact, *Aphis nerii* typically freezes in late autumn on milkweed stems, perishing on the plant, having had large populations, only to be an ecological and evolutionary dead end.

What determines the temporal peaks in the other milkweed insect species

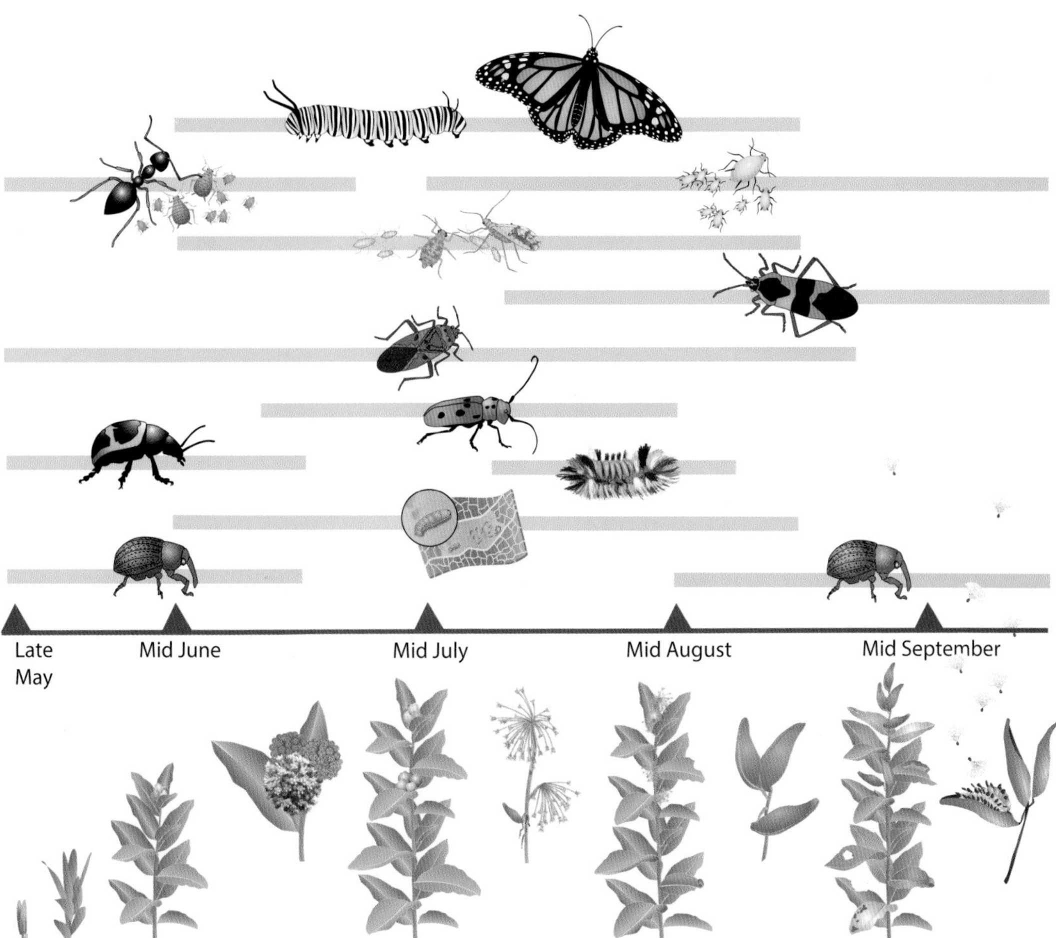

FIGURE 7.7. Timetable for eleven insects that feed on common milkweed in central New York. Note that the black weevil at the bottom of the diagram is, as far as we know, one species, although it feeds on stems in the spring and seedpods in the autumn. Organisms are not drawn to scale.

is less clear. For example, the two other aphid species peak early and later in the summer, respectively. For the first arriving milkweed insects, perhaps their strategy is to colonize clean and otherwise less-defended plants. Those insects that peak midseason, however, are environmentally less challenged, as both plants and insects typically enjoy the most rapid growth in those warm summer months. No matter what the cause, this established phenological order allows for predictions about which species should strongly interact.

An experiment conducted by a former postdoctoral researcher in my laboratory, Peter Van Zandt, now an assistant professor at Birmingham-Southern College, sought to determine the impact of early-season stem feeding by *Rhyssomatus lineaticollis* weevils on the assembly of the insect community later in the season. Pete's especially clever experiment allowed us to address specificity of induced defenses: are weevils special in how they induce defenses of the plants, or is it simply the fact that they are there early in the season? Of course, as a good scientist, Pete had plants that were protected controls (no weevil damage), those that he randomly assigned to a weevil attack treatment, and plants that were naturally colonized by weevils. Comparing the latter two groups is especially useful because it allowed us to contrast the effects of his controlled trial with how the natural system behaves. In addition, Pete wanted to compare the effect of weevil attack, which typically occurs early in the season, with that of monarch leaf feeding, which typically occurs much later in the summer. Because we kept a laboratory colony of monarchs, we were able to impose the monarch treatment at the same time as the weevil treatment, hence addressing the difference between the types of attack (not timing of attack) by these two species.

Weevil and monarch attack of milkweed had highly distinct impacts on the insects that later colonized. Weevil attack caused plant changes that reduced later feeding by aphids and monarchs, but attracted brethren weevils, probably weevils looking for mates. These effects induced by weevils were strong and lasted through much of the season. Monarch herbivory had more subtle effects, attracting *Tetraopes*, but these effects dissipated through the season. Therefore, there is a high level of specificity; not only is the timing of herbivory important, but the kinds of herbivores that colonize the plants directly affect the specific plant reactions. The specific herbivory-induced plant responses then set the plant on a trajectory for particular types of damage later in the summer.

Additional complexity derives from the fact that different plant parts may have different traits. For example, latex does not flow in the roots of milkweed, even though the roots contain the anatomical structures that carry latex elsewhere in the plant. The particular cardenolides that occur in roots, leaves, and phloem highly overlap, but not completely. Accordingly, many complex interactions may play out, and we are still working toward a predictive theory for understanding how specific chemicals, and where they are expressed, shape the communities of insects on plants.

A particularly striking example of asymmetry in such species interactions comes from recent work of a postdoctoral researcher in my laboratory, Jared Ali. Jared is a charming fellow and one of the world's foremost experts on soil-dwelling nematodes. While in my laboratory, however, I attempted to divert him, if only temporarily, into the wonderful world of monarchs and milkweeds. Jared and I tried to publish our findings under the title "Aphids suck, and monarchs rule"—and although the study was accepted for publication, the editor insisted we change the title to something less uncouth. Nonetheless, we still thought it was appropriate. Let me explain why. As you know, aphids are suckers; they suck phloem sap as their food. Among the many impacts that this has on the plant, one common response is to enhance plant quality for leaf-chewing insects, such as monarchs. There are likely many reasons for this, but one major explanation is that most plants seem to experience a hormonal trade-off in their ability to defend against aphids and caterpillars. Indeed, Jared found that monarch caterpillar growth was enhanced on plants previously fed on by *Aphis nerii*. In this same interaction, however, if monarchs arrive at the plant first and begin chewing, then aphid populations were suppressed. Aphids are susceptible to many different plant defenses, including those induced by chewing insects. Thus, among aphid-monarch interactions, it appears that monarchs always benefit, either directly by enhanced growth or by suppressing aphids, who eat the same resource. As suckers, aphids live under monarch rule. But, there is even more complexity, as could be anticipated from the differences between the aphid species that feed on milkweed.

So far in our discussion, all indirect interactions between insects on milkweed have occurred through changes in their shared host plant. However, there are other ways for these insects to interact, most notably through a fourth species, in this case an ant. *Aphis asclepiadis* is the only one of the three milkweed aphids that is nearly always tended by ants when growing in the field. Another postdoctoral fellow in my lab, Kailen Mooney, now a tenured professor at the University of California–Irvine, set out to test if aphid-ant interactions could affect monarchs. His idea was that if ants were tending to aphids on plants, and the ants ate, scared, or otherwise molested monarchs, then monarchs and aphids may be hitched together because the aphids attract predatory ants. This is exactly what he found. Milkweed plants with aphid colonies tended by more ants are less likely to have surviving monarch caterpillars.

So there are varied ways in which the milkweed insects are sewn together in a community. And milkweeds are certainly not unique. The communities of insects on plants typically interact in somewhat subtle ways. They are not usually driven by a limitation of green leaves but rather by the changing quality of that resource and the intersecting members of other species. Early-feeding insects induce plant changes that affect others. Ants and other insects are attracted to some milkweed herbivores, but once on the plant, their presence can influence other species. Together, these species represent food webs, not linear food chains or species pairs engaged in an arms race. Even so, we have shown that there are common threads, and together the milkweed insects have converged on many similar ways to exploit this toxic plant.

ONE FINAL MEMBER OF MILKWEED'S COMMUNITY

Homo sapiens have found many uses for milkweed. The plant has long been used medicinally. Thomas Edison showed that milkweed's latex could be used to make rubber. The oil pressed from the seed has some industrial applications as a lubricant, and perhaps even some value in the kitchen and as a skin balm. As a specialty item, acclaimed for its hypoallergenic fibers, milkweed coma (the

FIGURE 7.8. Milkweed seed pods opening in October near Ithaca, New York. The seeds and seed hairs, technically referred to as "coma" currently enjoy minor commercial uses. During World War II, American schoolchildren collected some 25 million pounds of pods, which were used to fill life vests with coma for the US Navy.

seed fluff that carries milkweed seeds in the wind) is currently used to stuff pillows and comforters (fig. 7.8). Perhaps more surprising, the same fluff is highly flammable and is used by survivalists to start fires. It is also so highly absorbent of oils, that it is now being sold in kits to clean up oil tanker spills. The fibers from milkweed stems make excellent rope and were used by Native Americans for thousands of years. More than two hundred years ago, the French were using milkweed fibers to make "cloth and velvets more lustrous than silk." This is a diverse plant with a lot to offer.

But thus far, most of what I have written in this book would likely (hopefully!) deter you from feeding on milkweed as a food. Native Americans and early French Canadians did collect nectar from milkweed flowers and use it as a sweetener, as milkweed nectar is mostly sugar and water. Yet for a common and abundant plant, what can one to do to make the vegetation palatable as a food resource? We humans, with our highly sensitive sodium pumps, do the one thing that milkweed insects don't do. We cook. In fact, the invention of cooking foods has been deemed one of the greatest advances in human evolution. Cooking certainly reduces the time spent chewing and digesting. But, perhaps more important, cooking opens up much of the botanical world—toxic as it may be—for human consumption, because heat breaks down many toxins, and hot water can also be used to leach out poisons.

Euell Gibbons, author of the 1962 classic, *Stalking the Wild Asparagus*, and famed proponent of wild plant edibles, was a huge advocate of eating milkweed. In his article titled "How to Milk a Milkweed," he wrote:

> The young shoots can be gathered in late spring when they are from four to eight inches high. Rub the natal wool off them, then cook and serve like asparagus. . . . If you get there too late for the shoots, don't despair. Gather the tender, young, top leaves and prepare them like spinach. . . . When flower buds that appear in the axils of upper leaves become greyish-green hemispheres an inch or so across composed of crowded, bead-like buds, they can be gathered and cooked in the same manner as broccoli. . . . Finally, when the warty pods are only about two inches long, firm in texture, and the silk and seeds are still underdeveloped, they can be gathered and cooked like okra. . . . All four of these vegetables are bitter—so bitter that few people can enjoy them unprocessed. . . . Fortunately, the processing to tame that bitterness down to palatable levels is easily done.

Gibbons recommended pouring boiling water over the vegetables in a pot, then heating only to regain the boil, and pouring off the water before sautéing. From my own experience, rinsing the vegetables with cold water several times tames the bitterness and preserves the burst of flavor produced during sautéing (fig. 7.9). "Season with salt, pepper and butter. Serve proudly."

At the end of summer, many herbivores have enjoyed the benefits of eating milkweed. Together, as a community, they have imposed natural selection on the milkweed and are joined together in a coevolutionary arms race. But while

FIGURE 7.9. Milkweed shoots collected in May 2015 from my front yard. I consider it a great achievement to have tricked my wife and son into believing that they were eating asparagus.

all others die off or go into diapause anticipating winter, monarchs are just preparing for an epic journey. Although the first summer generations of monarchs each live for about one month, the final "Methuselah generation" will live for eight months. The transition begins with the signs of autumn and a migration of thousands of miles by an insect that weighs less than a dollar bill.

CHAPTER 8

The Autumn Migration

I know not where their journey wends
What verdant land their prize
Perhaps 'tis where the rainbow ends
Perchance 'tis Paradise

—Valerie Dohren, "The Migration"

Fred and Nora Urquhart had been tagging butterflies for nearly three decades, and they had inspired and enlisted scores of citizen scientists to participate in their hunt for the monarch overwintering grounds. In January 1972, they had received a letter from a citizen scientist in Mexico who had seen thousands of roosting monarchs in the mountains near Mexico City. What's more, two butterflies tagged in the north were recovered, proving that a long-distance migration occurred. It was a hundred-year-old mystery, and they had worked to solve it for more than thirty years. And yet, they could not have imagined how impressive a sight it is.

There in the mountains of central Mexico, not thousands, but hundreds of millions of monarch butterflies go to roost and overwinter. And those individual butterflies have traveled thousands of miles to get there. It is a magical and tranquil place. Monarchs rest there for some four months, surviving mostly by staying cool and drinking a bit of water. When they huddle together, trees are literally weighed down with perhaps greater than eighty pounds (forty kilograms) of butterflies per tree (fig. 8.1). When they fly from their roosting trees, the whoosh of enumerable butterflies flapping feels like the gentlest touch of a feather. Their bodies are like glitter in a snow globe.

FIGURE 8.1. Clusters of monarch butterflies at an overwintering site in the Monarch Butterfly Biosphere Reserve.

This chapter is about how they get there. I will start by discussing what we know about the monarch's physiology, timing of departure, and orientation of flight at the end of each summer. Then we will turn to where they go, why, and the story of how it was found. Once monarchs begin their autumn migration in August, milkweed has mostly senesced, and it will not play a role in the monarch's biology for six months, until March of the following year, when the same butterflies remigrate to the southern United States.

AUTUMNAL CHANGES

In most plants and animals, seasonal changes bring about physiological changes. These fluctuations range dramatically, from the reabsorption of nutrients and shedding of leaves in deciduous trees (similar to the dieback of all aboveground tissues in milkweeds), to putting on fat in anticipation of hibernation in bears. Dogs seasonally shed and replace fur, while aphids and water fleas use seasonal cues to switch between giving live birth to laying eggs that will survive the winter. Seasonal shifts bring about serious challenges for organisms. And because the temperature and daylight cues associated with seasonal changes have persisted for time immemorial, natural selection has tuned organisms to use these cues.

Recall that by the end of summer, monarchs have had up to four generations of breeding since remigrating from Mexico. Each generation undergoes mating, egg-laying, and the battle between monarchs and the defenses of milkweed. But autumn conveys two major biological changes for monarchs in the northern United States and Canada: reproductive diapause and southern migration. In the life of a typical summer monarch during the first three generations, butterflies emerge from chrysalises ready to mate, with reproductive organs quickly maturing owing to the presence of "juvenile hormone." Juvenile hormone modulates many important insect functions, but is a bit of a misnomer for its role in monarchs because when it is in high abundance, insects proceed to reproductive maturity. As these summer monarchs emerge, mature, and mate during the summer, they do not employ directional flight, but rather they typically stay in the general locality where they had been caterpillars. Yet, beginning in mid-August, when the last generation of butterflies emerge, they have low levels of juvenile hormone, and this suppresses reproductive maturity. The result is that this last summer generation of monarch butterflies remain as virgins, migrate south, and wait months to mate.

Monarchs, like most organisms, pay close attention to a suite of environmental variables associated with seasonal change, including temperature, precipitation, and the number of sunlight hours (day length). As you might imagine, the most reliable and eternal of these cues is day length. Day length predictably increases after the winter solstice and decreases after the summer solstice. Although temperature and precipitation certainly change seasonally on average, there is much more year-to-year variation in the schedule of these changes than in day length.

Liz Goehring, a graduate student at the University of Minnesota, showed that three signals of oncoming autumn—the shortening of daylight hours (which begins after June 21 each year), large temperature swings between day and night, and declining plant quality—all contribute to reproductive diapause in monarchs. It is these same processes that also cause physiological changes resulting in increased fat storage and longevity of the butterflies. This autumn migrating generation has been dubbed the "Methuselah generation," named after the longest living person portrayed in the Old Testament. These butterflies will live some eight months, while the summer generations typically live only about a month each. Therefore, environmental cues trigger hormonal changes, like the lack of juvenile hormone, which ultimately lead to reproductive diapause.

Suppressing reproductive maturity and storing fat for travel and overwintering is only part of the story of the monarchs' long-distance migration. Soaring and gliding, orientation, and directional flight complete their autumnal changes. This southern flight, however, is not triggered by the physiological changes associated with the lack of juvenile hormone. The absence of juvenile hormone results in reproductive delay but does not affect directional flight. Thus, it is a different set of mechanisms that facilitate the monarchs' epic flight and how they are able to navigate toward their target, thousands of miles away in Mexico.

SOARING TO NEW HEIGHTS

The means of monarch migratory flight are spectacular. Individuals of the eastern North American population, which migrate the farthest, are large among monarch populations and, in particular, have larger front wings compared with nonmigratory monarchs. Like their migratory avian counterparts, monarchs fly high and take advantage of lift (upward sloping wind), thermals (pockets of rising warm air), and tailwinds. It is a special set of circumstances that allows an animal weighing less than a dollar bill to make this journey, taking more than two months, and typically more than 1,800 miles (3,000 km). With musculature fueled only by stored energy and by water and sugar from flower nectar, these butterflies can fly nearly a quarter mile high (1,250 meters above the ground). Over several decades, David Gibo of the University of Toronto studied many aspects of monarch flight through observations on the ground and in glider airplanes. Monarchs often ride thermals, columns of upward moving warm air, to great heights, and then glide with the push of any winds blowing in a southwesterly direction (fig. 8.2).

Monarchs also become semi-social during their southern migration, taking advantage of group dynamics to increase flight efficiency. In 1979, Gibo noted: "Team flying was common among butterflies soaring in thermals and appearing to increase the efficiency of the group in locating the stronger sections of the thermal. . . . On favourable days we were able to observe the formation of groups of two or three individuals. One butterfly would begin to circle and climb in lift and adjacent butterflies would quickly alter their course, using power when necessary, to join the higher individual. The group would then more or less circle in formation and climb until they were lost to view."

FIGURE 8.2. Monarchs' means of soaring on rising warm air (or thermals, blue arrows), which then allows for gliding down (or traveling long distances without powered flight). Monarchs can enter the thermals low to the ground or higher up. In either case, after entering, a circling and centering behavior facilitates lift. Monarchs often enter and ride thermals in small groups.

After watching hundreds of migrating butterflies in flight, Gibo estimated maximum flight speeds of thirty miles per hour (fifty kilometers per hour). In addition, he calculated that on a fixed amount of energy (140 mg of fat), a monarch, if moving by powered flapping only, could travel only eleven hours, after which its energy would be spent. Yet, by taking advantage of soaring and gliding, a monarch could theoretically fly ninety-three times longer (more than one thousand hours), reaching a distance of 6,200 miles (10,000 kilometers). Despite the stunning distance between the US Northeast and the overwintering grounds, Gibo concluded that stored energy was not limiting for this continental migration. It is time and tailwinds that are limiting, as even at fifty miles (eighty kilometers) per day, it would take the monarchs more than a month to complete their journey to Mexico. It is a race against the onset of winter.

a

FIGURE 8.3. (a) Southward migrating monarchs in autumn. Although monarchs are usually dispersed in the summer, as the autumn migration takes hold, flying butterflies congregate in larger clusters. (b) They also spend nights during the migration clustered on trees. If conditions are poor, migrating monarchs may stay clustered on

Monarchs have many challenges along the way. They must leave on time, collect floral nectar as fuel, and avoid too many delays en route. Variable autumn weather conditions often get in the way. They only fly during the day, and if there are headwinds (from the south), they often stay put or minimally use powered flight near the ground. Butterflies are constrained to minimize use of powered flight because it is expensive in terms of burning energy. Furthermore, unlike long-distance bird migrants, monarchs cannot make long glides without a major loss of elevation. Southward migrating butterflies roost in trees in the evening (fig. 8.3), and frost can kill monarchs in September, espe-

b

trees for a few days before continuing their journey to Mexico. Both images were taken by a citizen scientist on August 30, 2012, in Oshawa, Ontario.

cially those that have not made it to adulthood. Indeed, time is of the essence for these butterflies. If they do not commence the southern migration by mid- to late September, they will likely not make it to Mexico.

A MAP AND A COMPASS

In the autumn, as monarchs ready themselves for long-distance flight, their reproductive status stays virgin and they activate their navigation system. Over the past twenty years, major discoveries have been made about how monarchs plot their course. As articulated by Steven Reppert of the University of Massachusetts Medical School, one of the leading behavioral geneticists of our time: "To navigate across thousands of miles to a precise overwintering location the butterflies are likely to need a map and a compass, as used by other long-distance migrants." The "map" means that the animal knows where it currently is and which way to go to reach its destination, while the compass implies that the animal is able to maintain a more or less constant bearing toward that destination. In part because we still do not fully understand the monarch's map, I will first discuss its compass. What we now know is that monarchs use a "time-compensated sun compass" as a major part of their biological navigation system. As we will see, the critical initial experiments were published in 1997, and definitive additional work came along five years later.

Before its discovery in monarchs, the time-compensated sun compass was shown to be used by birds and bees, as demonstrated in classic behavioral experiments beginning in the 1950s. Using the sun makes sense, as it consistently rises in the east and sets in the west. However, the time of day is important, as merely flying toward the sun would send the animal one way at sunrise and in the opposite direction at sunset. Using the sun alone for directionality is not sufficient for making an effective compass. The "time-compensated" part of the sun compass allows the animal to use its perception of the time of day to adjust how it uses the sun to find a particular direction. An animal's sense of time is based on an internal "circadian" clock. Accordingly, an animal trying to fly south would orient such that the sun would be on its left in the morning and to its right in the evening (fig. 8.4, top panel).

Beginning in the 1950s, clock shifting (also called phase shifting) experiments were introduced to test the theory of a time-compensated sun compass. In such experiments, animals are initially trained to have their biological circadian clock to be out of phase with the natural sunrise-sunset cycle. Importantly, all animals are exposed to the same length of sunlight hours (say twelve hours of light) in a windowless room, but the artificial light would either start at midnight (six hours phase-advanced compared with actual sunrise), at 6 a.m. (the "control" group, in sync with actual sunrise), or at noon (six hours phase-delayed). After being in a phase-shifted environment for a few days, the animals' clocks adjust. All animals are still on a twenty-four-hour biological clock, but given that they use sunrise and sunset to calibrate their clocks, the phase-advanced animals now perceive the actual time of 6 a.m. as noon (because they have experienced six hours of light at this time). Phase-delayed animals perceive 6 a.m. as midnight (because they are not expecting sun for another six hours).

A very simple, elegant, and testable hypothesis was in place for animal migration using a sun compass. If the animals use their perception of the time of day to adjust their orientation based on the sun's position, then when placed back into natural light, at noon (12 p.m.), the control group attempting to fly south would fly toward the sun (fig. 8.4, middle butterfly in both the top and bottom panels). However, the phase-delayed group (perceiving it to be sunrise at 12 p.m.) would be expected to fly ninety degrees off the mark, to the west, at 12 p.m. (fig. 8.4, left side of bottom panel). On the contrary, the phase-advanced group, perceiving 12 p.m. to be sunset, would fly with the sun on their right, and thus would be heading due east (shifted by ninety degrees). Indeed, experiments have confirmed this use of the circadian clock to adjust flight orientation based on the sun in birds, bees, and monarch butterflies.

Given the use of such sun compasses in birds and bees, perhaps it is not surprising that monarchs use them too—although for such a small insect traveling such long distances, the convergent evolution of this navigational means across such diverse forms of life is remarkable. Perhaps more surprising was the discovery, which came in 2009 from Reppert's laboratory, that the circadian clock of monarchs is jointly calibrated by the brain and antennae. More than fifty years earlier, Fred Urquhart suggested that the antennae play an important role in the monarch's directional flight. Reppert's team used a diverse array of approaches, from molecular genetic analyses to releasing butterflies and watching which way they fly, to study the monarch's directed flight. Their key insights, however, were from relatively straightforward experiments,

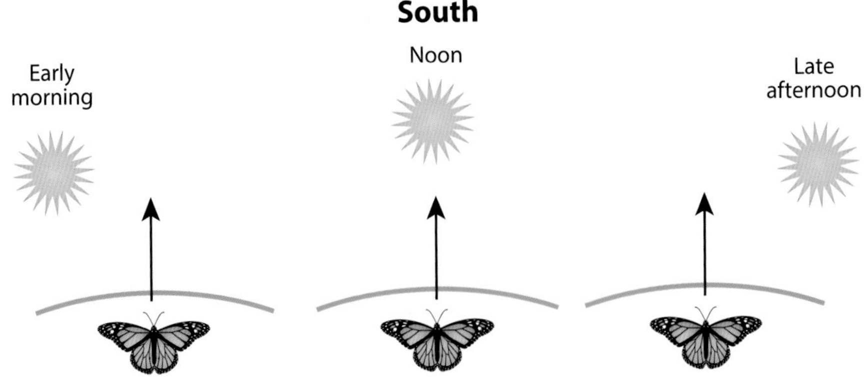

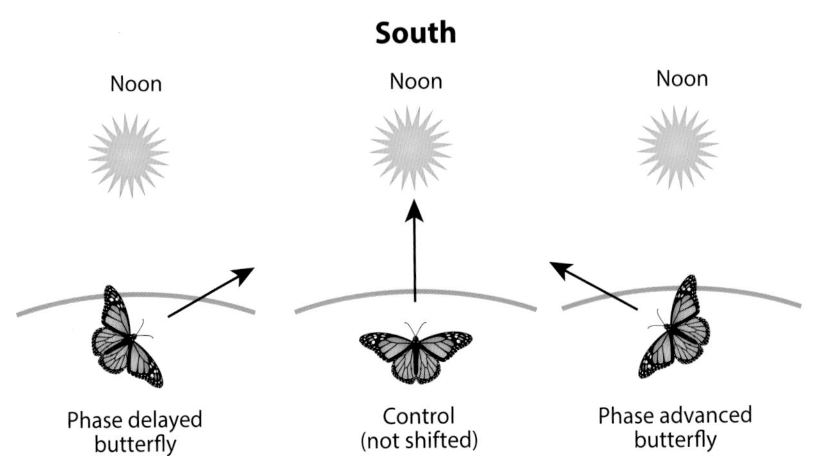

FIGURE 8.4. How monarch butterflies use their biological sun compass to fly toward Mexico. In the top panel, typical migrating monarch butterflies orient southward, independent of the time of day. They use their circadian clocks to adjust where they expect the sun to be, allowing them to use the sun as a "time-compensated" compass. In the lower panel, experimentally "clock-shifted" butterflies orient toward the wrong direction (not south) when tested at one time of day (see text for explanation).

where monarch antennae were removed or painted black, thereby removing solar cues to the antennal clock. Indeed, the antennal clock was critical to sensing sunlight as means to sense time and clock-set the compass. Migration takes more than a map and a compass, but also a clock to use the compass effectively. And the brain and antennae house the monarch's compass and clock.

BUT WHAT ABOUT THE MAP?

To migrate effectively, an organism needs to know where it is and where it wants go. Hence the need for a map. The mystery of how any long-distance migratory animal has this positional information is largely unresolved. However, a long-standing candidate has been the earth's magnetic fields, which emanate from the earth's core and eventually meet charged particles coming from the sun. These fields are distinctly aligned at the north and south pole, like a giant magnet. Therefore, the magnetic field generates a predictable gradient across the earth's surface. And if animals have a compound called magnetite in their bodies, as monarchs do, then they may be able to use this field for positional information. Loggerhead sea turtles and pied flycatcher birds, which both make tremendous long-distance migrations, have been well-studied for their use of magnetic maps to make major decisions along their route. In particular, a set of extraordinary experiments showed that when these animals were put into large magnetic fields simulating different geographical regions, the sea turtles and flycatchers would predictably reorient. Their map is magnetic, but what about monarchs?

Although there have been claims and speculation for decades, the role of the earth's magnetic field for monarch mapping is only now being revealed. Monarchs were recently described to use magnetic information, but primarily as an additional mechanism to their sun compass. In other words, monarchs use magnetic fields to maintain directional flight, which is especially important on cloudy days. However, it is so far unknown whether they use magnetic fields for positional information (as a map). Nonetheless, monarchs must have a map of some kind—they cannot "learn" the necessary positional information because the butterflies flying south in the autumn are several generations younger than those that flew north in the spring. From studies of tagged butterflies that have subsequently been recaptured well before they reach Mexico, it is clear that monarchs direct their flight toward their target (fig. 8.5). Monarchs migrating from the Midwest tend to migrate directly south, while those from the northeast tend to fly in a southwesterly direction. Once monarchs reach Texas, they are funneled further south. If they are diverted east, they are bounded by the Gulf of Mexico, and toward the west by the Rocky Mountains (and later the Sierra Madre Orientals). Such visual cues are very coarse, however, and may only help to direct butterflies that are badly off course.

Another way that scientists have addressed whether monarchs "know where they are" is through geographical displacement studies. In these cases, migrating monarchs have been captured and moved hundreds to thousands of kilometers away, and their flight directions have been followed. Here, their existing sun compass could continue to help them orient in particular directions, but a "map" would be required to know what direction to migrate in order to reach Mexico. Such experiments have been attempted several times, and none have revealed clear evidence for a map sense that is detectable by magnetic fields. In the earliest studies, Fred Urquhart displaced monarchs from Ontario to several western provinces and states, more than a thousand kilometers west. However, when released, the monarchs continued to fly southwest, as they would to reach Mexico if they were still in southern Ontario. Such experiments were repeated twice beginning in the 1990s, but again, it appeared that migrating monarchs displaced either east or west of their collection location continued to migrate as if they were at the original location. It is unknown if these butterflies eventually would have corrected their course or if they were unable to reorient. The butterflies were observed for only a few hundred meters until they vanished from view of the investigators.

I speculate that the migratory trajectory and map may be determined during larval development, from the time a caterpillar begins to grow, up to its emer-

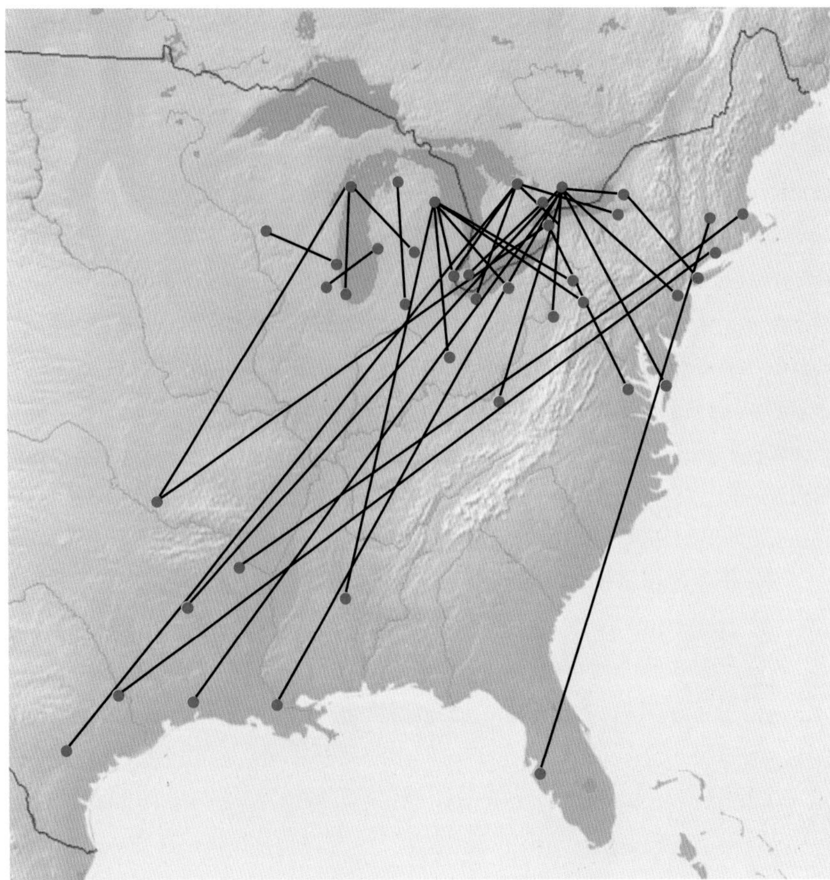

FIGURE 8.5. A map showing "release-recovery lines" produced by Fred Urquhart in 1965, showing northern locations where monarchs were tagged by citizen scientists and southwestern locations where they were recaptured. Each tag contains a unique identifier.

gence as an adult butterfly. In other words, perhaps the magnetite takes time to align in their bodies and is only fixed, setting the map in place, over the course of larval development. In this case, short-term displacement studies of adult butterflies may not be sufficient. What is needed is to take monarch eggs from a single location, to rear them from egg to adulthood in different locations, and then to examine their orientation when released. If their orientations differ from adults that are shipped to different locations, but reared in one place, this would be good evidence that the migratory trajectory is determined during the caterpillar's development. This experiment is yet to be conducted.

Overall, the monarch's map sense is still a mystery. Although monarchs contain magnetite in their bodies and use the earth's magnetic field to orient, there is so far little evidence that they use the magnetic field as a map in the way that some sea turtles and birds do. Interestingly, the magnetic charge in the mountains of Michoacán, Mexico, close to where the monarchs overwinter, is known to be anomalous and shows very high magnetic readings. Perhaps the butterflies are attracted to this emanation. But so far, this is only conjecture. Migrating monarchs certainly use visual cues as well, such as major bodies of water, mountains, and river valleys. It is generally agreed that these cues may be useful in funneling butterflies in particular directions, or to get them back on course if displaced, but these cues are unlikely to be driving the monarch migration to specific mountain peaks in Mexico.

Before we leave this topic, it is worth noting that the map and compass associated with this seasonally induced migration is genetically programmed in butterflies. It could not be "learned" in any traditional sense, as each of the previous monarch generations was either in a different place, or if in the same place, was not migratory. Nonetheless, the butterfly's genetic program may allow it to assess the place of birth and environmental cues associated with that location (including magnetic gradients). Monarchs possess the genetic machinery to make a map and to use a compass, but the execution of these migratory tools are implemented with local information. It takes four to five generations annually to complete the monarch's migratory cycle. But in one of those generations, not only do changes occur in the physiology, morphology, and brain, but a journey of unimaginable lengths takes place.

WHITHER GOEST THOU?

The monarch overwintering sites were certainly known for hundreds of years to native peoples who lived in the central Mexican highlands. Nonetheless, these were somewhat isolated communities, and there is little historical information about the cultural significance of monarchs in Mexico. The native Purépecha people celebrate the Day of the Dead (Día de los muertos) in early November, coinciding with the arrival of monarchs to the region. And legend has it that the butterflies are the returning souls of loved ones. Despite this, it is still unclear why the monarch overwintering sites did not assume greater cultural significance historically, as the sheer numbers and clustered nature of the butterflies on trees would have been expected to make a strong impression on the local residents. The great Mexican author Homero Aridjis spent his childhood in the 1940s in the region, and he climbed the hills near the town of Contepec, where at least one of the large colonies of monarchs used to overwinter (this area has since been logged). His childhood observations stimulated his writing, including the poem "To a Monarch Butterfly," published in his 1971 book *El poeta niño* (see the epigraph to chapter 1).

As discussed in the first chapter, although there were suspicions of a southern migration of monarchs, scientists did not know where they went. In 1870, the *American Entomologist and Botanist*, a naturalist's periodical, published a letter to the editor on the subject of monarch migration:

> During my ramble this morning I happened upon a flock or bevy of butterflies, known as *Danais archippus*, Fabr., containing thirty individuals. . . . I find them to be . . . identical in every respect with specimens bred from the caterpillar by myself last summer, except in that of color, which is somewhat paler in these captured this morning than it was in those bred by me in the summer. They have the appearance of having been on the wing some days. The interesting question is, do they hibernate in the [adult] state, or in that of the chrysalis? They are wholly in advance of their larval food-plant, *Asclepias obtusifolia*. Please give us the facts as to the manner and condition in which they spend the winter, and oblige yours, respectfully, L. J. Stroop. Waxahachie, Texas, March 31, 1870.

The editor, Charles V. Riley, one of the leading entomologists of the late 1800s, replied: "They undoubtedly hibernate in the perfect [adult] state, for we have often captured pale, faded and worn specimens quite early in the spring of the year." But, he did not include any information about where they might have spent the winter.

Despite this spot-on insight, debate among entomologists about whether they did "hibernate," and if so, where, would continue for nearly one hundred years. Beginning in the mid- to late 1960s, Fred Urquhart was convinced that monarchs flew into Mexico and flew north from Mexico in the spring based on returned tags from the citizen scientists. But he still did not know what they were doing there. In his 1966 newsletter, he wrote: "One of the BIG mysteries of the monarch migrations is, where do they go from Mexico? We suspect that they travel along the Gulf Coast of Mexico as far as Central America and perhaps reach the Pacific coast there to mingle with the southern extension of the California population. But, we have very little data to support such conclusions. Hence, we are bending every effort to tag as many butterflies in southern Texas as possible. We also hope to supply our associates in Texas with many hundreds of butterflies next fall." Soon after, Urquhart had a sabbatical leave, which he wrote about in May 1968: "As a result of my years of service to the University of Toronto, I have been granted a sabbatical leave of absence. We, Mrs. Urquhart and I, will be joining our colleagues at the Texas A & I University in Kingsville, Texas, next winter [January 1969]. We hope to carry on our studies there in a location that will permit an on-the-spot investigation of the movement of the monarchs along the Gulf Coast, into Mexico and perhaps to the Pacific Coast."

In 1970, Urquhart began to call for increased involvement of research associates in Mexico and Central America, with the annual newsletter advocating the involvement of friends and colleagues in tagging as well as requesting aid in translating Spanish. The hunt for the overwintering grounds was in full force, with momentum building. In the 1973 annual newsletter he wrote about their trail of discovery. Urquhart was rightfully proud of their scientific journey, and he recounted hypotheses that had been erected and how they rejected each of them over the past decades. Monarchs do not hibernate under logs, as had once been suggested. Adults do not spend the winter flying freely in Florida and the Gulf Coast. And monarchs do not fly around the southeast to Texas, up to California, and back east again in the spring. Laboratory experiments,

FIGURE 8.6. The cover of the August 1976 issue of *National Geographic* magazine, announcing to the world the discovery of the monarch overwintering sites. Featured on the cover at the Cerro Pelón colony is Cathy Brugger, one of the key citizen scientists who helped to locate the colonies.

tagging of butterflies, and tireless field expeditions demonstrated each of these past hypotheses to be false. At last, with the help of thousands of citizen scientists, the Urquharts now had reports in hand of overwintering butterflies in Mexico. He wrote: "This is indeed a most amazing flight and one which, a few years ago, would have been considered quite impossible for an insect to accomplish annually."

Still, it was another two years before the massive overwintering colonies were found and reported to the world. After thirty years of tagging, intense research, and much speculation, in 1975, the great discovery was made. "We now wish to announce to our associates, that, after these many years of intensive study, after having tagged thousands of migrants, we have, finally located the exact area where they overwinter, with the very able assistance of Ken Brugger and Cathy Brugger of Mexico City" (fig. 8.6).

FIGURES 8.7. The monarch overwintering grounds in central Mexico in the Monarch Butterfly Biosphere Reserve.

Photographs simply do not do justice to the monarch overwintering colonies. Although the density on covered trees is difficult to fathom (up to five thousand butterflies per square meter from ground to treetops), there is a feeling of something altogether different when experiencing the quiet fluttering of thousands of butterflies that have come to roost from thousands of miles away. The largest colonies have up to 300 million butterflies (fig. 8.7). As Robert Michael Pyle, the founder of the Xerces Society for Invertebrate Conservation and author of *Chasing Monarchs: Migrating with the Butterflies of Passage*, described his monarch overwintering experience in the *New York Times*, "A gold curtain parts and a rain of golden sequins falls before you. You enter a forest chamber where every surface—ground, rocks and trees—is covered by butterflies. It is a world of butterflies filling the air as you are entirely enveloped by golden wings."

My first experience at the overwintering sites took place just after New Year's Day in 2012. A friend and colleague, Karina Boege, a professor at the National Autonomous University of Mexico, and Eduardo Rendón-Salinas of the World Wildlife Fund Mexico, led me and my family to the colonies at the Cerro Pelón Sanctuary and El Rosario. The confluence of several attributes make the monarch overwintering colonies very special. Hiking in at more than ten thousand feet (three thousand meters) above sea level sets the stage, as the air is thin and the sun is intense, even in winter. The butterfly sanctuaries themselves appear as forested islands on the mountaintops, with a sea of surrounding agricultural fields. Nights are cool, mornings are filled with quiet fog, and the Oyamel fir trees are weighed down, each with tens of thousands of butterflies (fig. 8.8).

THE URQUHART-BROWER BROUHAHA

The discovery of the monarch overwintering sites was an international sensation. Butterfly enthusiasts, nature-gawkers, and others were dumbfounded by the images published in *National Geographic* magazine. At the time, Lincoln

FIGURE 8.8. Oyamel fir trees at El Rosario (Monarch Butterfly Biosphere Reserve). Note that the trees on the left are weighed down with dark clusters of butterflies. Individuals of the same tree species, not occupied by butterflies, are seen on the right side.

Brower, one of the early pioneers of monarch biology, was eager to get to the overwintering sites to further study their biology. In particular, he wished to study the intersection of their larval feeding biology, sequestration of cardenolides from milkweed, and protection from predation. Brower had personally observed enormous clusters of overwintering monarchs in California, and he reasoned that they might be hundreds of times larger in Mexico. Not only a wonderful sight, but an unprecedented opportunity to study predation on this toxic icon.

FIGURE 8.9. Lincoln Brower (in green sweater) confronting Fred Urquhart (in leather jacket and cap) on January 22, 1977, at the Sierra Chincua site of overwintering monarchs.

Given the annual reports emerging from Urquhart's research, in 1973, Brower contacted Urquhart by letter, asking if they had found any overwintering sites. Brower's request was met with enthusiasm and a response that indicated that Urquhart would provide the information when available. Nonetheless, despite Brower's persistent letters and phone calls, Urquhart ultimately refused to reveal the locations, even after the publication of the *National Geographic* article in August 1976. Anxious and annoyed, in particular about the lack of collegiality, Brower doubled his efforts to locate the colonies on his own. Teaming up with a young lepidopterist, William Calvert, they used two key pieces of information presented in the *National Geographic* article: the sites were in northern Michoacán state, Mexico, and they were mountaintops around ten thousand feet above sea level. Led by Calvert, on January 22, 1977, Brower arrived at the overwintering colonies at Sierra Chincua. He reported: "As fate would have it, we encountered the Urquharts and Bruggers tagging monarchs inside the colony. The Urquharts were bewildered by our arrival and initially treated us rudely, and then with hostility" (fig. 8.9).

FIGURE 8.10. Satellite image of the monarch overwintering grounds in central Mexico. Although up to thirty colony areas have been found, thirteen areas are indicated by orange circles as the ones most consistently occupied (#4 has only occasionally been colonized in recent years). Four sites (#5, 6, 10, and 11) are the largest and most consistent. Names are as follows: (1) Mil Cumbres, (2) San Andres, (3) Altamirano, (4) El Cedral, (5) Sierra Chincua, (6) El Rosario, (7) La Mesa, (8) Chivati-Huacal, (9) Lomas de Aparicio, (10) Cerro Pelón, (11) Piedra Herrada, (12) Oxtotilpan, (13) Las Palomas.

Although some rancor was precipitated by the encounter, and the *New York Times* published an article about their rivalry, the biologists both continued their research across the entire range of monarchs from southern Canada to Mexico (fig. 8.10). As we will see below, the discovery of the colonies led to some wonderful insights about their overwintering biology, especially the impact of birds and mice as predators. The discovery of the overwintering sites also initiated a persistent focus on the conservation of monarchs.

SUCH A SNOWBIRD

What we now know is that the monarchs end up congregating in a tiny area, with the bulk of the butterflies concentrated among twelve mountain massifs (clusters of peaks) within three hundred square miles (eight hundred square kilometers), an area smaller than New York City. In other words, most of the monarchs from eastern North America, from Maine to Saskatchewan, and south to Texas, probably covering two million square miles, funnel down and overwinter in a location 0.015 percent the area that they occupy in the summer. Although some thirty sites have been documented with overwintering colonies, the largest and most consistent sites number about a dozen. The location of these colonies is somewhat predictable, and clearly very special. In some years, monarchs return to the specific trees that the population roosted in the previous year, while in other years, they occupy nearby trees. Still, there is no evidence of any sort of specific site fidelity. All of the monarchs that funnel to these mountaintops are a mixture of butterflies from across the United States east of the Rocky Mountains, and there is no return to particular colony sites based on their ancestry. How and why individuals end up at a particular overwintering colony is a mystery.

Monarchs preferentially overwinter on Oyamel fir trees (*Abies religiosa*), a species that is highly restricted to the cool climates of the high elevations in central Mexico. In fact, the species is reminiscent of Canadian fir trees of the boreal forest (also in the genus *Abies*), and is limited to only a handful of moun-

a

b

FIGURE 8.11. Aerial view of monarch overwintering habitats: (a) Cerro Pelón (near where the first overwintering sites were discovered in 1975). Unforested meadow areas called *llanos* are often nearby the high elevation Oyamel fir forests and may play a role in monarch orientation toward overwintering sites. (b) Monarchs cluster on Oyamel firs at the Rosario colony, about twenty kilometers away from Cerro Pelón as the crow flies. Note that the orange color of the trees next to the *llanos* are concentrated clusters of butterflies.

taintop "islands" in Mexico, typically around ten thousand feet above sea level. Monarch colonies are usually found on somewhat steep southwest-facing slopes (fig. 8.11). The ultimate reasons why monarchs overwinter as part of their life cycle and the specific reasons why they do it in the highly restricted Oyamel fir forests are manifold.

WHY MIGRATE?

Monarchs are epic migrators, but why any animal journeys in such a way, even for shorter distances, is a complex question. Perhaps the first thing to note is the specific meaning of migration itself. Although there are many definitions, the one most useful in this case is the seasonal movement of animals between two regions, where conditions are reciprocally favorable and unfavorable, typically with breeding occurring in only one of the two regions. Many animals, from dragonflies to hummingbirds to wildebeest make annual migrations. The evolution of such migrations is thought to have had its origin in large populations that are compelled to take advantage of a geographically available resource.

Insects, including monarchs, however, are fundamentally different from most of their vertebrate counterparts in that individual vertebrate animals typically make the round-trip. Yet as already discussed, monarchs (and most other insects) produce multiple generations between their initial migration and the return trip. This allows for greater time to migrate and less energy required per individual, but it adds the problem of orientation and movement toward areas never before seen by the individual animal. A few attributes have likely contributed to the monarch's migratory behavior: its evolutionary origin in the tropics, the progressive seasonal availability of milkweeds moving northward, and the ability of the butterfly to fly long distances, navigate across generations, and survive it all, especially the defenses of multiple milkweed species.

Monarch butterflies are part of a tropical group (the Danaini, introduced in chapter 1). What this means is that the common ancestor of all the Danaini, a single butterfly species that eventually gave rise to the 170 species we have today, lived in a tropical environment and was likely unable to survive freezing temperatures. Some animals have certainly evolved from tropical ancestors and adapted to the temperate zone with freezing tolerance. Even butterflies in the same family as monarchs, the Nymphalidae, spend the winter in the temperate zone as a quiescent caterpillar or adult (for example, the Viceroy butterfly overwinters as a caterpillar). But this is not the case for monarchs. In fact, given their tropical ancestry, monarchs likely originally fed on Mexican milkweeds and simply avoided freezing temperatures all together.

However, as with most species, the availability of a new and high-quality resource is often strong motivation to move and colonize that resource. For migration to evolve, moving to consume some resource must enhance fitness over staying local. And that resource for monarchs is milkweed. Mexico has a wonderful diversity of milkweeds (more than fifty species of *Asclepias*), yet most are rare in the landscape, and as the season progresses, there are fewer plants available and the leaves decline in food quality for caterpillars. Further north, milkweeds sprout in the spring in the southern United States and continue to sprout later in the summer as the season progresses northward. Thus, following newly sprouting milkweeds makes evolutionary sense, in particular by allowing additional butterfly generations to feed on the progressively emerging milkweeds.

There may be other benefits of migration as well. As was described in chapter 6, monarchs have several enemies, including parasitic microbes and insects. Especially for small and mostly sedentary parasites, the migration has long been hypothesized to keep parasite loads low. Indeed, in migratory populations of monarchs, like that of eastern North America, parasite loads are consistently lower than in nonmigratory populations. Because migrating butterflies move from place to place, parasite populations never build up. In addition, infected individuals are often removed (or "culled") from the population because they are not able to complete the journey. Thus, benefits of migration include two mechanisms of reduced parasite infection, both of which could support the evolution of this strategy.

WHY HERE?

Although the monarch has a tropical ancestor, through its expansion and exploitation of northern milkweeds, the insect has settled on a largely temperate lifestyle. The bulk of monarch mating, reproducing, and feeding is conducted in the United States. But conditions favoring these behaviors are not critical in the overwintering grounds. What is needed at the overwintering site is the right environment to bide their time through the months that it is snowy in the north. Although there are many reasons that monarchs overwinter on their particular mountaintops, one key to understanding their choice of sites appears to be climate. Given that monarchs cannot tolerate freezing, the lower bound of their temperature requirements is obvious; it should typically stay above freezing. Importantly, the upper bound on comfortable temperatures is also critical. If too warm, the butterflies become more active and must adopt cooling mechanisms, which use up lipid stores needed to persist over the four months of overwintering. Draw-down of lipid stores is a key factor in monarch mortality at the overwintering grounds, and as mentioned in chapter 4, many that survive but do not have enough lipid reserves stay in Mexico, returning to the lowlands and breeding on local milkweeds. The narrow climatic band needed by monarchs is well represented in the high elevation forests in Mexico. Additionally, although there are very few flowers providing nectar during these winter months, monarchs are able to drink essential water from morning dew droplets and streams nearby.

The second attribute for the perfect overwintering retreat would be that it is safe from predators. On the one hand, given that monarchs sequester cardenolides from their milkweed host plants, one would expect predation to not be an issue in any overwintering area. On the other hand, such a large congregation of any "resource" would surely attract predators that could learn or evolve to eat that resource. Given that overwintering monarchs are greater than 30 percent lipids (which is high-energy food for predators), they are often quiescent on trees or the ground, and they are so dense as to weigh down trees, they should be a highly sought-after resource. Predation was one of the first things that Brower and Calvert investigated at the overwintering grounds beginning in 1977.

Calvert, and later a graduate student, Linda Fink (who would eventually marry Lincoln Brower), studied bird predation by two species in particular, the black-backed oriole (*Icterus abeillei*) and black-headed grosbeak (*Pheucticus melanocephalus*). These birds were reported to eat hundreds of thousands, if not millions, of butterflies each year—up to 60 percent predation in some colonies. Bird predation is greater on the outside (or periphery) of the monarch colonies, and is inversely proportional to colony size (higher predation in smaller butterfly colonies). Calvert and Brower reasoned that these relationships may favor large, dense overwintering aggregations. Still, two interconnected questions needed answering to explain this unexpected phenomenon of high bird predation. First, how do the birds do it? And relatedly, why aren't the cardenolides effective?

Orioles are indeed sensitive to cardenolides, and these birds can taste the toxins. Accordingly, orioles pin monarchs down with their feet and use their beaks to slice open the abdomens and thorax. By slicing open the bodies, orioles are able to consume the contents without ingesting the cuticle, the latter of which has much higher concentrations of cardenolides. After opening up monarch bodies, the birds sometimes discard their catch. Fink and Brower reported that the greater the cardenolide contents of the butterfly, the less likely the oriole is to eat the body contents. Orioles also preferentially consume male over female monarch butterflies, likely because males typically contain about one-third less cardenolides than females. Grosbeaks, however, use an entirely different tactic. After ripping off the monarch's wings, they eat the entire body (fig. 8.12). What Fink and Brower showed next was truly marvelous.

In forced-feeding experiments, they demonstrated that orioles showed the typical vomiting response to eating cardenolide-laden monarch bodies. Orioles are indeed sensitive to cardenolides, use dexterity to reduce their expo-

FIGURE 8.12. Two ways to eat monarchs: (a) Black-backed orioles pin a monarch and cut open the abdomen to consume the contents; they leave the exoskeleton and wings. (b) Black-headed grosbeaks typically clip monarch wings and then eat the rest of the body whole.

sure, and use taste to accept only the most palatable individuals. Grosbeaks, however, did not show the typical vomiting response. They are more resistant to cardenolides. Grosbeaks have an unknown physiological means—either not uptaking the cardenolides into their bloodstream, perhaps having insensitive sodium pumps, or even shuttling the toxins away from the most sensitive areas. Even so, grosbeaks showed a seven-day cycle of eating many monarchs and then taking a break, suggesting that they may also sicken, recover, and then resume eating. Although the specific physiological means have not been studied in either bird species, the connection between their feeding behavior and physiology is a promising avenue of research.

Mice also gorge on monarchs at the overwintering sites. Although several species of mice eat monarchs, only one, the black-eared mouse, *Peromyscus melanotis*, makes monarchs a big part of its diet. When most other mice species are fed only monarchs, they lose weight because they simply refuse to eat very many. The black-eared mouse, however, chomps them, even eating enough to allow the mice to breed in the wintertime, which is rather unusual for mice. Not unlike the orioles described above, the black-eared mouse primarily feeds on abdominal contents, discarding the cuticle, and prefers to feed on lower-cardenolide butterflies over those that have sequestered more toxins. As is true for the orioles, mice also preferentially eat male monarchs. Through a series of experiments and observations over several years, John Glendinning, a doctoral student with Brower, concluded that the black-eared mouse was the primary mammalian predator of monarchs at the overwintering sites because it was the least sensitive to the bitter taste of cardenolides. Although all of the mice species tested were somewhat negatively affected by the milkweed toxins physiologically, only the black-eared mouse was not deterred by the bitter taste, and this was the only species to slice open the abdomens to harvest the lower-cardenolide internal tissues. Some birds and mice have clearly gained behavioral means to reduce cardenolide exposure and can apparently tolerate low doses of cardenolides better than other species.

In the end, are the overwintering grounds really a safe haven? The mountaintops, and the Oyamel fir trees in particular, certainly provide an ideal climate for the monarchs to rest and wait out the winter. And the large, dense, and toxic clusters of butterflies provide safety in numbers. Birds and mice have a field day and consume many butterflies, but apparently not enough to offset the benefits of multiple generations afforded by the migration. Because milkweed is available for most the spring and summer in the United States, the population can rebound and grow. Even if there are catastrophic die-offs owing to high rates of predation and occasional weather anomalies, migration and overwintering are still highly successful strategies.

Many researchers had previously held the belief that migration appeared late in the monarch's evolutionary history, perhaps as recently as the past several hundred years. On the contrary, recent genetic evidence suggests that migration was an ancestral condition in the monarch's evolutionary lineage. It seems that migration may be deeply imbedded in the monarch's genome and has the potential to continue to evolve. For example, monarch populations that budded off the eastern North American population, such as those in California and Australia, have developed alternative migratory routes and likely use alternative environmental cues. There are nonmigratory populations as well, such as those in south Florida, Spain, and Hawaii, and it appears that these have evolutionarily lost migration. Still, there is a lot we do not know about the genetic basis of migration. For example, it is unclear whether butterflies, if translocated from nonmigratory populations to eastern North America, would regain migration in autumn simply through environmental cues.

The spectacular multigenerational annual cycle of monarchs in eastern North America holds many mysteries, promises, and problems. Although some of the mysteries of monarchs' flight orientation have been solved, others relating to their "map sense" remain. Current research on monarch butterflies continues to reveal novel insights into neurobiology, physiology, and ecology. A problem, however, of sustaining an annual cycle that crosses the boundaries of three huge nations is that any break in the cycle could be catastrophic. Since the discovery of the overwintering grounds, monarch biologists, starting with both Urquhart and Brower, have signaled concerns about the population's long-term viability. In the next chapter, I will address what is known and not known about the sustainability of monarch butterflies.

CHAPTER 9

Long Live the Monarchy!

Through the night, coated in frost,
the woods around my town wait for the light of dawn.
Like closed leaves, the monarch butterflies
cover the trunk and branches of the trees.
Superimposed, one upon the other, like a single organism.
The sky goes blue with cold. The first rays of sun
touch the clusters of numb butterflies
and one bunch falls, opening into wings.
Another cluster is lit and through the effect of the light
splinters into a thousand flying bodies.
The eight o'clock sun opens up a secret that slept
perched on the trunks of the trees,
and there is a breeze of wings, rivers of butterflies in the air.
Visible through the bushes, the souls of the dead
can be felt with the eye and hand.
It is noon. In the perfect silence, the sound
of a chainsaw is heard advancing toward us,
shearing wings and felling trees. Man, with his thousand
naked and hungry children, comes howling his needs
and shoving fistfuls of butterflies into his mouth.
The angel says nothing.

—Homero Aridjis, "About Angels IX"

The environmental movement began to take hold in the decades preceding the discovery of the monarchs' overwintering grounds in 1975. Prior to this, monarchs were simply a quiet icon of nature. It was the beauty behind metamorphosis from caterpillar to butterfly, the science of mimicry and sequestration of toxins, and the mystery behind their migration that captured people's attention. Nonetheless, milestones of the era included the publication of Rachel Carson's *Silent Spring* in 1960, and Richard Nixon's double boon of creating the Environmental Protection Agency in 1970 and advancing the Endangered Species Act of 1973. Increased environmental awareness and rapid advances in knowledge of the monarch's biology produced what some have dubbed "the Bambi of the insect world"—perhaps the most recognized butterfly on the planet and now an icon of conservation.

Ever since the announcement of the monarchs' overwintering grounds in Mexico in 1976, the conservation of the species and its migration have been at the forefront of scientific studies as well as the public's understanding of science. A 1977 article in the *New York Times* that described the rivalry between Fred Urquhart and Lincoln Brower noted: "Both scientists, who were careful not to disclose the exact location of the overwintering site in their respective articles, agreed on one thing: The need to protect the butterflies and the environment that nurtures them." There are several reasons why conservation became an issue after the discovery, most notably that without understanding their basic ecology and annual migratory cycle, it had been difficult to specify the critical limiting factors for monarch populations. With knowledge of the highly specific nature of their overwintering sites, ecological interactions with milkweed, and their long-distance migration, scientists could now more rigorously study their population dynamics.

Over the past forty years, several common themes related to monarch conservation emerged. As for any species, its habitat is its home, and without a healthy habitat the species cannot thrive. For the monarch, much of North America is its habitat, and this is both good and bad news. Good because there is seemingly endless expansive space, but bad because the completion of the migratory cycle requires utilization of much of that space, even if only for brief periods each year as the butterflies move from location to location. In terms of the overwintering stage, the bulk of butterflies east of the Rocky Mountains congregate on tiny mountaintop islands in Mexico for several months, making protection of that specific area a persistent concern. Its habitat in the United States and Canada is expansive, but its overwintering habitat in Mexico is miniscule.

Outside of habitat protection, other threats relate to the key ecological needs of any species. As was made clear in earlier chapters, we all need to eat and avoid being eaten. The monarchs' exclusive food source for larval development is milkweed. Without milkweeds, there are no monarchs. And as we will see, the declining number of milkweed plants is currently one of the most popular explanations for reported declines in the size of the monarch population. Although monarchs have evolved remarkable strategies to avoid being eaten, the litany of vertebrate, insect, and microbial predators and parasites could, if unchecked, be a threat to their sustainability as well. As outlined in chapter 6, a huge fraction (often more than 90 percent) of caterpillars fall as prey. Additional human activities, ranging from the overexploitation of resources, to the use of herbicides and insecticides, and the introduction of invasive species have been suggested as threats to the monarch, either through their direct impacts, or through negative effects on the butterflies' habitat, food, or safety from predators and parasites.

IS THERE REALLY A PROBLEM?

I have a friend who is an environmental sociologist at Cornell, and his take on the state of the planet and its environmental future is that things are far worse than we can ever imagine. The environment is rapidly declining, and we are, by and large, in denial. An alternative view, one held by many Americans, is that there are environmental problems, but they are solvable, and that science and technology will aid in their solution. This dichotomy represents a range of what people believe. What people believe is not necessarily easy to evaluate, especially using the scientific method. The state of the planet, precisely when we will reach particular tipping points, and how close to extinction a given species may be is extraordinarily difficult to know. As an endeavor, "conservation" involves some facts, many unknowns, and a set of beliefs. These beliefs typically involve both assumptions and speculation. Conservation is both a natural as well as social science, formulated with a large dose of beliefs. As such, with certain facts in place, assumptions must be made, costs and benefits are weighed, and ultimately some action may be taken.

Here the initial set of facts are fairly clear cut. There is indeed a problem, and the numbers of monarch butterflies are certainly declining. Although there has been much speculation about this over the decades, the clearest evidence for a population decline came in 2012, when Lincoln Brower and colleagues published an analysis of the monarch population size at the overwintering grounds since 1994. Yearly censuses have been taken in a standardized way in Mexico, allowing for an analysis of the temporal trend. The trend observed by Brower has more or less continued (fig. 9.1). Monarchs in western North America, concentrated in California, have also been declining, although the extent to which this is independent or caused by the same mechanism at work in the eastern population is unclear. And although this story is about monarchs, it should also be frankly noted that many of the world's great migrations, including those of mammals and birds, are declining and equally threatened.

While in the early 1990s there were some 400 million butterflies roosting each winter in Mexico, in the past few years, the number has hovered around 100 million butterflies, or a 75 percent decline (based on the average of the early 1990s versus the average of the last few years). In February 2016, as I write this chapter, news was just released that this winter there were some 200 million monarch butterflies in Mexico (well over triple the number last year). Still, any way you slice it, there has been a serious and apparently persistent long-term decline.

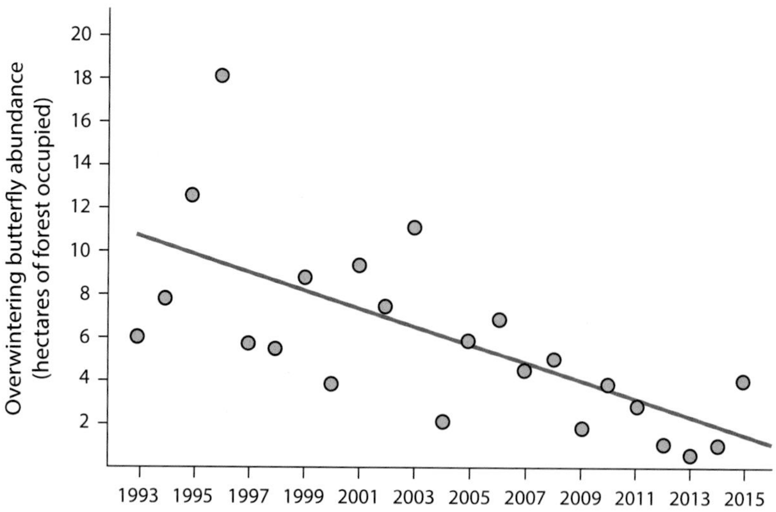

FIGURE 9.1. Monarch butterfly abundance in the forest roosts at the overwintering grounds in high-elevation mountains of central Mexico. Because it would be impossible to count all of the (hundreds of millions) butterflies, the area of forest occupied by dense colonies is surveyed. Approximately 50 million butterflies occupy a hectare (about 2.5 acres) of forest with dense colonies. This estimate is for the population that remigrates back to eastern North America (the California population is somewhat independent). Note that the overwintering season starts in November and ends in March; the year on the *x* axis indicates the beginning year of each overwintering season. In other words, 2015 indicates overwintering 2015–16.

And yet, in the same year that Brower published this analysis of the population size in Mexico, Andrew Davis, of the University of Georgia, published an analysis of two other annual monarch censuses of south-flying butterflies at the beginning of the autumn migration. These fall censuses are conducted in a different way than those conducted at the overwintering grounds. Let me explain. As monarchs fly south in the autumn, they often aggregate and fly through "funnel points," especially next to large bodies of water. As such, several peninsular sites along the Great Lakes and the Atlantic Coast serve as concentrated flyways for monarch butterflies (and also migrating birds). With tremendous foresight, a standardized protocol to monitor autumn migrating butterflies was established in the early 1990s at Peninsula Point, Michigan; Cape May, New Jersey; and Long Point, Ontario. Observers count butterflies

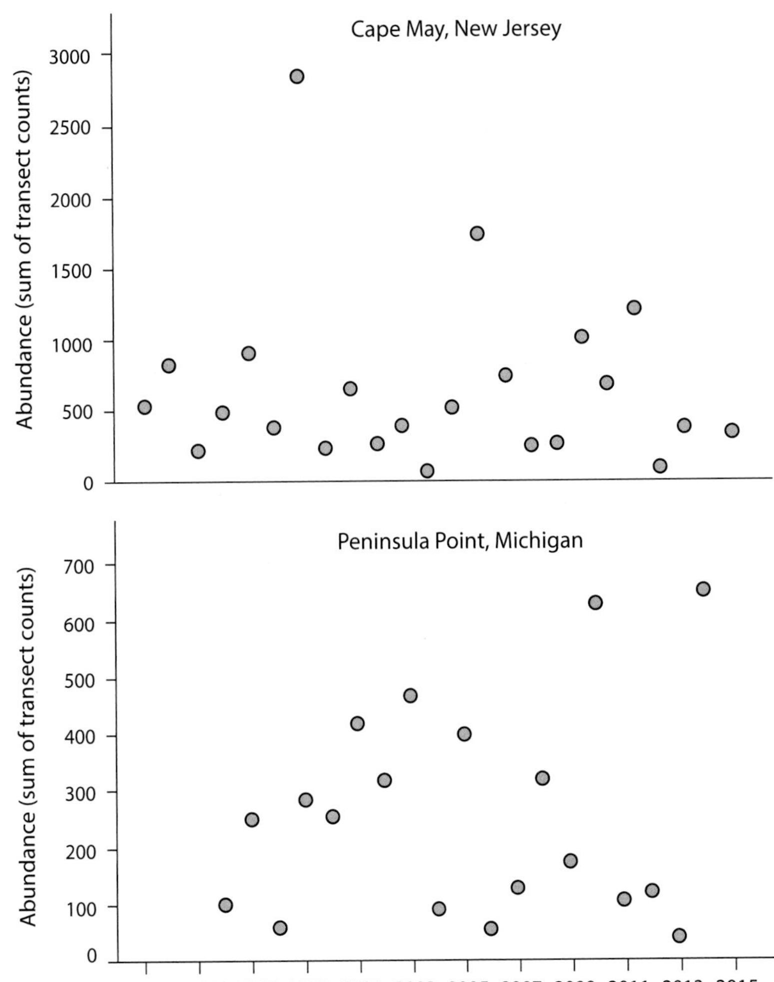

FIGURE 9.2. Estimates of the annual monarch population migrating south through two peninsular funnel points in autumn. Note the lack of a declining population over time. Also, the highest number of recorded monarchs flying through Peninsula Point was in 2014.

as they move along a defined transect several times per day during the entire autumn migration period. Analyzing these data, Davis found no decline of monarch populations over the past two decades (fig. 9.2).

The discrepancy between what Brower and Davis found piqued my interest. I brought it up for discussion at the Monarch Biology and Conservation Conference in June 2012. Much to my surprise, some monarch biologists thought it was inappropriate to discuss the discrepancy, because, to them, there was no ambiguity in the decline. One senior monarch researcher grabbed my hand, and chastised me for speaking about variation in monarch declines. However, other monarch biologists welcomed the discussion, and shared uncertainty as to the extent of the monarch's decline and its possible causes.

When I returned home, I set out to acquire funding to address the biological problem of discrepancy in the various census data (Brower's versus Davis's conclusions outlined above) and to address the social issue of how individuals and organizations use scientific information as well as the iconic nature of the monarch as a symbol for conservation. This intersection of natural and social science is precisely at the nexus of what conservation is, the coming together of uncertain facts and social dynamics. This was also the type of interdisciplinary project being advocated by a new academic center at Cornell, the Atkinson Center for a Sustainable Future, which provided a grant for this work. I teamed up with two Cornell colleagues, Steven Wolf, a professor of environmental sociology, and Bruce Lewenstein, a professor of science communication. Together we hired a talented graduate student studying population dynamics, Hidetoshi Inamine, and a postdoctoral scholar in sociology, Karin Gustafsson. What ensued was an incredible *Sturm und Drang* of discovery and distress.

One of the early results that emerged from our research was that the monarch butterfly has historically had wildly fluctuating populations year-to-year. Take the winter seasons 2002–5 plotted in figure 9.1, above. In those years, monarch populations increased 50 percent (2002–3), declined 80 percent (2003–4), and then increased by 200 percent (2004–5). What a roller coaster of population dynamics. Fred and Nora Urquhart's first synthesis of the two hundred or so citizen scientists' observations in 1955 already gave important information about such population fluctuations. In 1953, monarchs were exceedingly rare, almost absent throughout the breeding grounds in midwestern and eastern North America, whereas they were plentiful in the previous two years. Ann Swengel, a citizen entomologist and monarch enthusiast published an analysis of summer butterfly counts taken by citizen scientists throughout the midwestern and northeastern United States between 1977 and 1994. In this case, butterfly counts are taken around the country and reported to the North American Butterfly Association. Here too considerable variation was found in year-to-year monarch numbers, with the population estimates indicating a quadrupling increase or crash over the course of a few years. And the three recent years (2013–15) have been a similar roller coaster. After reaching the all-time low in 2013, the population doubled the next year, and in 2015 it more than tripled compared with 2014. What could cause such wild fluctuations?

POPULATION BOOMS AND BUSTS

We know that the weather, in particular the timing and amount of precipitation, as well as temperature, have a considerable impact on monarch populations. Some of these effects may be direct, such as mortality caused by snowstorms at the overwintering grounds, which will be discussed below. In addition, much of the effect of precipitation occurs via indirect effects on milkweed. In years with abundant, lush-growing plants, especially in the spring in the southern United States, monarchs tend to be in greater abundance later in the summer. To be sure, the devastating drought in Texas between 2010 and 2014 (the worst drought in fifty to one hundred years) must have had a negative impact on monarchs. But perhaps most critical to our understanding of monarch population fluctuations is the multigenerational aspect of the annual cycle. There is an up and a down side to these multiple generations: huge reproductive potential on the one hand, and the reliance of each generation on the previous one, on the other hand.

Recall that the last generation of butterflies in the summer migrates to the overwintering grounds in autumn, and these same butterflies migrate back over the border in spring to lay the first generation of eggs in the southern United States (see fig. 4.1). Following their larval development and metamorphosis into butterflies, that first generation migrates north to colonize the midwestern and northeastern states, where two to three additional generations occur. Let's examine the reproductive potential of a single female butterfly that successfully migrates from the north in the autumn, survives several months of overwintering in Mexico, mates, and flies to Texas. If she lays 300 eggs, all of which survive, and half of which are female, she will leave 150 successful progeny who themselves can lay eggs. Assuming three additional generations, the math is relatively simple: 150 females times 150 female progeny times 150 female progeny times 300 progeny (the total, including males and females in the last generation) equals just over one billion butterflies. A single monarch female can theoretically yield a billion descendants over one annual cycle.

The above example is absurd for a few reasons, most notably because we know that there is never 100 percent survival of eggs laid. On the contrary, most die in battle with plant defenses or at the hands of predators and parasites. Additionally, many females do not lay all of their eggs owing to the many hazards they face. Still, that a single butterfly could produce a billion offspring in one annual cycle is striking. Such exponential population growth is possible in all organisms, as was recognized early on by Thomas Malthus and Charles Darwin. Population biologists sometimes refer to this phenomenon as "multiplicative fitness," namely, the idea that the production at each generation may lead to a multiplying effect in the next generation. But this reliance of each generation on the past one is not necessarily a positive one for the overall size of the population.

Take for the moment the above scenario for the first three generations ($150 \times 150 \times 150 = 3.4$ million female butterflies). However, now imagine that there was little milkweed left for a fourth generation, so only one out of ten females was able leave a successful (single) female offspring. In this case, 3.4 million is multiplied by 0.1, leaving a mere 340,000 butterflies. If we imagine an even more dire circumstance, where a winter storm kills 75 percent of the remaining butterflies, the population would be reduced to less than 85,000. Given the tens of thousands of monarchs eaten by birds and mice, and the possibility of a poor autumn migration, one could easily imagine the monarch population to be in trouble. That is the nature of population dynamics, a potential for boom and a potential for bust. Even in introduced populations of monarchs, such as that in southern Spain, butterflies are often at low densities, but twice over the past decade they have had booms, or outbreaks, that have locally defoliated all milkweeds (which are also not native there).

So, is there really a problem? Yes. The consistent decline at the overwintering sites is cause for concern. The pattern is very robust and, on average, population sizes are currently about half that recorded two decades ago (see fig. 9.1). Because the monarch's annual cycle occurs over several generations, the population may be able to grow in some generations. Nonetheless, the annual decline through the "bottleneck" of the overwintering population is symptomatic of a major problem. A persistent decline in the face of large annual population fluctuations (much of which is due to weather) may eventually push monarchs to dangerously low numbers.

SUNDRY THREATS TO THE BELOVED MONARCH: LOGGING AND CLIMATE CHANGE

Direct physical damage to the highly restricted overwintering sites was an early and continued threat to the monarch population. The area is used for timber (as fuel and building materials), cattle grazing, and hunting. Logging has been a persistent issue at the overwintering sites, and this has direct effects on monarchs. Plain and simple, without the specific forest type needed to overwinter, monarchs cannot survive. In addition, even "thinning" or selective logging has strong effects on monarch viability, although these effects are largely indirect, by influencing the climate at a very local scale. Overall, monarchs prefer to establish roosting colonies in more mature "closed canopy" forests, in the middle of Oyamel fir trees that are taller than sixty feet (twenty meters). The middle of trees is preferred because this is where there is "thermal constancy"—it stays warmer at night and cooler during the day than closer

FIGURE 9.3. Monarch butterflies roost in the middle of these Oyamel fir trees (note the orange coloration on the lower branches, which are covered with monarch clusters, but not on the tree tops).

to the forest floor or toward the top of the canopy (fig. 9.3). As Lincoln Brower has pointed out, thick tree trunks act as hot water bottles, absorbing heat during the day and giving it off at night. Large and dense trees also act as critical "jackets" and "umbrellas," providing insulation and a barrier to wind and rain, both of which are problematic for monarch overwinter survival.

There have been several attempts to protect the overwintering grounds, from the initial decree by President Lopez-Portillo of Mexico, who in 1980 stated: "Because of public interest, those areas in which the butterfly known as the Monarch hibernates and reproduces are made a sanctuary and reserve." This first step did not designate boundaries or make a formal proposal for habitat conservation, and hence in 1983 the International Union for Conservation of Nature designated the monarch migration as a threatened phenomenon. Note its wording: "The species *Danaus plexippus*, indigenous to the New World and extensively established in Australasia, Hawaii and elsewhere, is not endangered. However, the phenomenon of the migration of the North American population may be seriously threatened. The main threat is from logging of the conifers in which the winter roosts occur."

In 1986, Mexican President Miguel de la Madrid established the 60-square mile (16,000-hectare) Monarch Butterfly Biosphere Reserve, which was expanded to 56,300 hectares in 2000 (217 square miles in total, with no logging allowed in the "core" 13,500 hectares, which covers 25 percent of the reserve). The site became a UNESCO World Heritage Site in 2008. A consistent challenge in protecting the overwintering grounds has been to create alternative income opportunities for local communities and to develop sustainable ecotourism. For example, the Monarch Fund was established in 1997 to compensate locals for loss of access to the reserve land and to concomitantly help achieve conservation goals.

Despite these efforts, the high elevation forests used by monarchs to overwinter have seen a steady rate of logging and degradation (fig. 9.4). In an important study that used historical aerial photographs, Brower and several colleagues from the World Wildlife Fund (WWF) and National Autonomous University of Mexico reported that between 1971 and 1999, 44 percent of the intact forest in and around the Monarch Butterfly Biosphere Reserve was degraded. The annual rate of degradation increased after 1984, suggesting that initial legislation to protect the overwintering grounds was ineffective. In a 2014 update, Omar Vidal and colleagues from the WWF reported that between 2001 and 2012, about 16 percent of remaining intact forest in the core area was deforested or degraded, nearly all by illegal logging. This logging was thought to have been halted in 2009, as there has been a marked increase in law enforcement, as well as continued investment in creating alternative economic opportunities for locals. Yet, in early 2015, several news articles reported the

FIGURE 9.4. Recent illegal logging in the Cerro Pelón Sanctuary in the Monarch Butterfly Biosphere Reserve (September 2014). Note the decaying mossy stumps in the foreground and background from logging in past decades.

loss of at least fifty acres (twenty hectares) of forest, that more than twenty-five loggers were arrested, and that six illegal sawmills were shut down in the region. Vidal and colleagues wrote: "Small-scale logging is a serious and growing concern for the conservation of the monarch sanctuaries and the reserve. Its dynamics need to be better understood so as to devise a strategy to curtail it. It was not until 2012, when we compared photographs from 2001 with those from 2011, that the considerable effect of this activity became evident" (fig. 9.5). At that time, some twenty-seven thousand people lived in farming communities within the buffer zone (although not in the core areas), and more

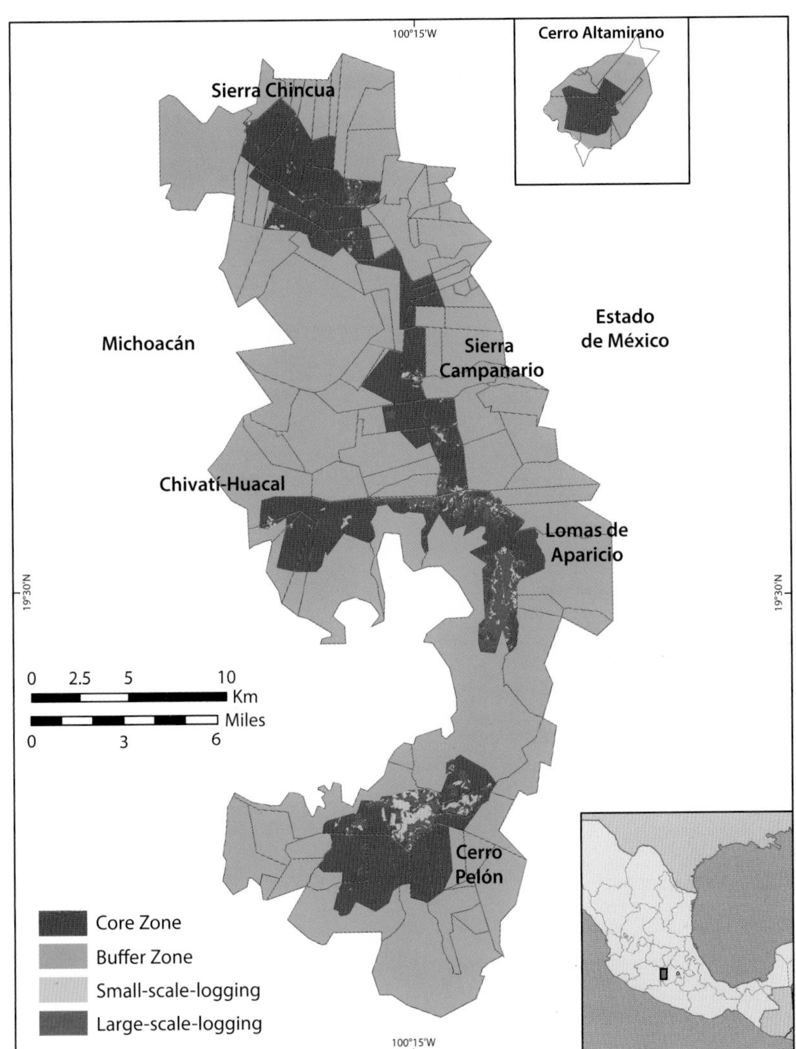

FIGURE 9.5. Map of the Monarch Butterfly Biosphere Reserve, with logging impacts shown in orange and red. Surrounding areas are mostly agricultural lands that are heavily used by local communities.

than one million people lived outside the reserve, many of whom were formerly employed in forestry and mining. Additional reports of substantial illegal logging have trickled in through March 2016 at the time of this writing. The extent of continued illegal logging in the Monarch Butterfly Biosphere Reserve is, at best, unclear.

The future of the Oyamel fir forests and their suitability for monarch overwintering depends not just on logging, but on climatic effects. Although the impacts of floods and fire are relatively minimal, climate change scenarios do not bode well for these forests. Recently developed models indicate that Oyamel habitat may shrink by as much as 70 percent in the next fifteen years and by greater than 95 percent over the next seventy-five years. Such dramatic shrinkage is likely because high-elevation tropical montane areas are among the most sensitive habitats to shifting climates. Severe alterations in both precipitation and temperature, the two main factors that restrict Oyamel fir habitat, are expected in the coming decades. Other models predict that monarchs do not currently inhabit all of the available overwintering habitat in the region, and that higher-elevation sites might become suitable in the future. However, such habitats are outside of the Biosphere Reserve, currently do not have Oyamel firs, and climate shifts are occurring far faster than the trees could establish in those areas. Thus, even if forests could be established in such locations, other variables such as poor soils and increased probabilities of winter storms could make the habitat unsuitable for monarchs to successfully overwinter.

WINTER KILLS

Monarch butterflies are amazingly resilient organisms, but they are, at heart, a tropical insect. When temperatures fall, butterflies can survive to 18 degrees Fahrenheit (−8 degrees Celsius) if dry, but only to about 23 degrees (−5 degrees C) if they are wet from precipitation. Since the overwintering grounds were discovered in 1975, there have been five major freezes that have killed millions of butterflies and several more storms that have been close calls. The causes and consequences of such winter kills have been the subject of intense discussion among monarch biologists. Climate change has been implicated, but it is difficult to pinpoint how frequent such events were in the past.

Even if winter storms typically occurred every decade or so, as has been observed since the discovery of the overwintering sites, it has been suggested

FIGURE 9.6. Massive mortality of monarchs at the Rosario site (Conejos colony) caused by a snow and ice storm in early March 2002. Shown is one of Lincoln Brower's undergraduate assistants sitting in a pile of dead butterflies more than twelve inches deep.

that the impacts of these storm events on the population have been magnified as the forests have become degraded. Winter storms in January and February, which typically include snow, rain, and temperatures just below freezing, have killed substantial fractions of monarchs (20–80 percent mortality) in 1981, 1992, 2001, 2002, and 2004 (fig. 9.6). Smaller, less devastating storms occurred in 1995, 2010, and 2016.

Orley "Chip" Taylor of the University of Kansas, who established Monarch Watch in 1992 as an outreach and education program focused on monarch butterfly biology and conservation, has written extensively about monarch overwintering and storms. Monarch Watch was the natural inheritor of the Urquharts' Insect Migration Association, which managed the tagging program and produced an annual report for citizen scientists. Monarch Watch similarly established a tagging program, as well other initiatives, and also produced annual reports, which later transformed into blog posts.

In February 2003, Taylor summarized the fluctuations and impacts of the droughts and storms over the preceding few years: "The monarch population has crashed and recovered in dramatic fashion several times in the last 4 years." He went on to summarize the greater than 60 percent decline from 1999 to 2001, which he attributed to a drought that affected plant availability (not milkweed, but other late-summer plants that provide nectar). To add insult to injury, late winter storms dramatically increased mortality, and "we had never seen such a low population, so the prospects for the next season did not look good." Nonetheless, the monarchs bounced back, more than tripling the population that next summer. Yet another severe winter storm the following year knocked the population down by 75 percent. Again the population bounced back fully. Taylor summarized: "Monarchs have been fortunate. Two bad winters have been followed by two reasonably good breeding seasons." Nonetheless, he warned that poor winter survival in Mexico could eventually be coupled with a dry breeding season in the north. "We haven't been tracking monarchs long enough to be able to estimate the likelihood of low overwintering populations being followed by harsh, unfavorable, summers, but this will happen; the question is when."

The above analysis has been the state-of-the-art knowledge about weather, winter storms, and monarch population fluctuations. It has been assumed that harsh weather in the northern breeding grounds and the overwintering areas is critical to population declines, but that as long as good seasons occur intermittently, population recovery is possible. As was seen in the early 2000s, the population crashed and recovered more than once. Nonetheless, over the past decade another factor has begun to create widespread concern: do monarch caterpillars have enough to eat?

THE MILKWEED LIMITATION HYPOTHESIS

As I write this, "not enough milkweed" is a pervasive and widely accepted view for the cause of the decline of the eastern North American monarch butterfly. It is in the news seemingly weekly. Monarch Watch and countless other organizations are selling milkweed seedlings and otherwise encouraging milkweed planting. A year ago, the President Obama's family even planted some milkweed in their "kitchen garden," which included some inedibles. First Lady Michelle Obama said, "A pollinator garden helps to encourage the production of bees and Monarch butterflies. They pollinate the plants, they help the plants grow." Although flowers and pollinators certainly rely on each other, as discussed in chapter 2, monarch butterflies themselves are rather poor pollinators, especially of milkweed. Are monarch populations truly limited by the abundance of milkweeds?

Although milkweed limitation has grabbed many recent headlines, it has been considered for at least forty years. By the late 1970s, the Urquharts began advancing the milkweed limitation hypothesis. They wrote in their 1979 newsletter, "The monarch butterfly is becoming a rare species in some parts of North America. Where once monarchs would be found feeding upon the nectar of flowers along roadways and pasture fields and, in the autumn, thousands would be seen winging their way to the overwintering site, only a few are now ever seen. This rather sad state of affairs is due primarily to the destruction of the milkweed plants as the result of rapid urbanization and the indiscriminate use of herbicides." In this same issue, milkweed seeds were offered for planting upon request to the Urquharts. Brower also worried about the agricultural use of herbicides and the lack of milkweed, as evidenced by comments he made in the *New York Times* in 1986 and 1998. Even so, this concern about lack of milkweed took a back seat to the woes of deforestation in Mexico until recently.

The milkweed limitation hypothesis is simple. First, the only food suitable for monarchs' development is milkweed. As has been detailed in the earlier chapters of this book, monarchs and milkweeds have coevolved, resulting in a highly specialized and dependent relationship from the monarch's perspective. Milkweed itself has no need for monarchs. Second, milkweeds and especially common milkweed (*Asclepias syriaca*) have been declining in the agricultural Midwest, largely as the result of the expansion of agriculture and use of herbicides. Herbicides kill weedy plants, and that is beneficial to crops, especially because weeds are the number one agent that reduces crop yields (more so than insect herbivores or microbial pathogens). Therefore, as herbicide use has increased, milkweed abundances have decreased. And this effect has become stronger since the deployment of genetically modified (GM) crops, which harbor a gene making them tolerant of herbicides. Herbicide tolerance is one of those feats of bio-engineering that allows farmers to clean their fields of weeds, even after the crop has begun to grow. The adoption of herbicide-tolerant corn and soybeans in the United States began around 1997 and has increased ever since. Today, more than 90 percent of the acreage of corn and soybeans is GM herbicide tolerant. The third part of the milkweed limitation hypothesis is that lower milkweed numbers translate into lower monarch numbers. This is a reasonable assumption, but as I will discuss below, not one that has been rigorously tested.

Despite the growing momentum of the milkweed limitation hypothesis, when I look around field sites in central New York State, my intuition tells me that milkweed is plentiful. Walking through the fields, I typically find a monarch caterpillar on less than 3 percent of the milkweed stems I encounter, unless they are resprouts from a mowed field, in which case there may be more. In either case, monarchs rarely defoliate plants, and there appears to be plenty of milkweed. But if milkweed scarcity is indeed limiting to monarchs, perhaps when patches of plants are found by butterflies, they will lay abundant eggs. Conversely, perhaps milkweed is not the limiting factor.

Over the course of a month in the summer of 2014, I went in search of milkweeds and milkweed habitat in the northeastern, southeastern, and midwestern United States. With my friend and fellow milkweed fanatic, Mark Fishbein, I circled Apalachicola National Forest in northern Florida, and drove from there to central Florida, finding twelve species of *Asclepias* along the way. I only found monarchs on *Asclepias perennis*. Then I traversed my home region in central New York, up to and beyond Lake Ontario and returning along the eastern coast. There was milkweed everywhere (I encountered six species), with monarchs present on three of them. Finally, I traveled to my father-in-law's home in central Illinois and found much milkweed, three common species but with very few monarchs. My family and I then took the train through the heart of the corn and soybean belt from Illinois, ending in Denver, Colorado. Here too were four common species of milkweed, and although other milkweed insects flourished, there were few monarchs. What I saw was ample milkweed, but very few monarch butterflies or caterpillars.

Although my intuition of plentiful milkweed was corroborated, there were very few monarchs to be seen. And the intuition of others about milkweed's decline was not necessarily wrong. In fact, as I passed through central Illinois, hosted by Bill Handel, a botanist with the Illinois Natural History Survey, it was impressive to see the expansion of agriculture, with crops often grown right up to the road edges, with little to no "margins" where weeds like milkweed would grow in the past. Still, many highway margins, which were no longer mowed every year, natural areas, state parks, conservation zones, and, yes, even waste places, had abundant milkweed. Could monarchs be declining because of partially marginalized milkweed, or is there another cause?

CONNECTING THE DOTS

With a problem of such enormous spatial and temporal scale, it is difficult to imagine how to test the milkweed limitation hypothesis. After all, our intuitions, whether based on observing abundant milkweed or observing the agricultural destruction of milkweed habitat, may not be sufficient. Nonetheless, because of the monarch's annual migratory cycle, which relies on milkweed in March through August, but not for the other six months each year, identifying transition points in the cycle at which monarchs decline would provide important information about a variety of factors, including milkweed limitation.

We tackled the issue by first trying to link the stages of the migratory cycle. Already in place were annual censuses of the overwintering grounds, the summer censuses across the United States reported to the North American Butterfly Association (NABA), and two censuses at the beginning of the autumn migration collected at two funnel points near water. All these censuses, despite being independently organized, have generated consistent and high-quality data beginning in the early 1990s. To verify the quality of the data and assess if monarch populations behave as we expected, we sought to link the successive stages of their migratory cycle. We asked if banner years of overwintering monarchs translated into large populations as the butterflies progressed northward. Similarly, we expected that in years with high or low summer numbers, the censuses during the autumn migration and the following winter in Mexico should follow suit. As you will see below, the fact that the censuses were conducted independently and using distinct methods (area occupied in Mexico, butterfly counts per hour of group observation by NABA, and the sum of migrating butterflies at point locations in the autumn) strengthened our conclusions.

There were several important results from this research. First, estimates of the overwintering monarch population in Mexico (based on area of forest occupied by the butterflies) strongly predicted the butterfly counts observed in the following spring in the Southern United States. This result makes sense, as it is the very same butterflies that migrate north from Mexico in the spring, lay eggs in the southern United States, and generate the first new generation of monarchs of the year. Second, we found that the number of butterflies observed in the southern states in spring predicted, although with less accuracy, the numbers observed in the midwestern and northeastern states. These are the second through fourth summer generations that rely on milkweed as a host plant (as did the first generation in the southern United States). And these populations, which build over the summer, predict (although with even less

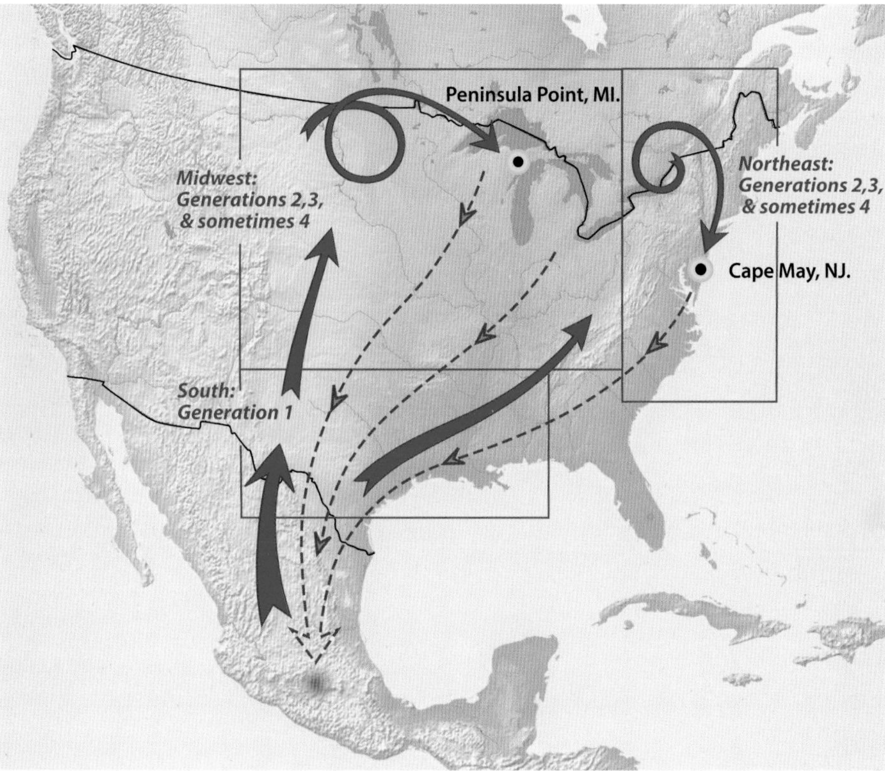

FIGURE 9.7. The annual migratory cycle of monarchs east of the Rocky Mountains. Solid arrows up from Mexico are drawn progressively thinner because the initial population grows, and by the end of the summer, the population size could not have been predicted from the starting population in Mexico. The dashed arrows representing the southern migration indicate a lack of correspondence between the number of butterflies starting the migration and the numbers that are ultimately found at the overwintering grounds in Mexico. The break in the cycle appears to be during the southern migration and establishment in the Monarch Butterfly Biosphere Reserve.

accuracy) the censuses at the start of the southern migration at the two funnel points. However, we found no connection between these end-of-summer population estimates and those that would then overwinter in Mexico. The break in the cycle appears to be the transition from newly minted adults in the north to successful migrants overwintering in Mexico (fig. 9.7). Critically, the break in the cycle occurs after the monarchs are done feeding on milkweed.

An additional analysis confirmed this result. When we accounted for the "input" (the population size at the previous stage) at any given stage north of Mexico, that input drove the population abundance observed. In other words, the population size in the Midwest was best explained by the number of migrants from the southern United States in the spring. Again, this makes intuitive sense. However, the only transition where this was not the case was the transition between autumn migrating butterflies and the estimates of the overwintering population. Here, a linear trend of monarch population decline was present over the years, regardless of the "input." In other words, each year, the starting population in Mexico predicts the next stage, and the next stage predicts the following one, and so on, although with decreasing predictive power at each step. This decrease in predictability is expected, as each generation allows the population to grow, but various environmental factors may allow it to grow more or less. A temporal population decline was evident in Mexico and the southern United States, but not in any of the northern censuses. Again, this is consistent with our other analysis, as the input was becoming less and less important, as the monarch population builds each generation as it moves north.

The remaining question is perhaps the most important one. Why are monarch populations declining in Mexico if they can build up each year during the summer? Why are they declining after they are through completing their generations on milkweed? Unfortunately, we don't know the answer. Yet.

DISCLAIMER

The community of biologists who have contributed substantially to our understanding of monarch biology is relatively small. Perhaps fewer than fifty living scientists. Most agree that there is indeed a problem with the sustainability of the monarch butterfly population that migrates across eastern North America each year. However, many of these scientists, several of whom have spent their professional lives studying the monarch, disagree with the conclusions I have presented above (that milkweed is not limiting). Many scientists believe that milkweed is limiting, and that this has caused the decline of monarchs. I will further evaluate the milkweed limitation hypothesis below, but my group's analysis indicates that the main problem with the monarch population occurs during the autumn migration, after monarchs have completed development on milkweed. Hidetoshi Inamine and I (along with our collaborators—his PhD adviser, Stephen Ellner, here at Cornell and Jim Springer, data wizard of the NABA) have applied the most rigorous analyses we can to the existing data. Our results have been controversial. And unfortunately, with such a large-scale problem, one of a population decline occurring over decades, and with a population that spans much of North America, we are stuck with the messy data we have in hand. New data can be collected, but little can be done to add to our current understanding of the population dynamics over the past forty years other than to work with the existing data. Nonetheless, I believe that the existing data have not yet been fully exploited, and thus there is hope to gain further insight from both past data and from continued monitoring projects. A report just issued by a blue-ribbon panel of the US National Academies of Sciences (May 2016) agrees that there is not enough evidence to support milkweed limitation as the cause of the monarch decline. I expect that further analyses and debate will help resolve the disagreement among scientists on this matter. So far we have heeded the many cautionary comments and letters sent to us, warning us that they strongly disagree with our conclusions, and these comments have pushed us to increase the rigor of our analysis.

DATA, CORRELATIONS, AND CAUSATION

Much of the story about the potential drivers of declining monarch populations has been based on intuition or observations like those made during my journeys in search of milkweed, but not on hard data. Over the decades, many explanations have appeared in print, including logging, severe weather and climate change, insecticides, and now loss of milkweed as a consequence of herbicide-tolerant GM crops. What is somewhat different about the current focus on loss of milkweed is that it is based on data, although as we will see below, the causal basis of the relationship is not as strong as it could be.

In 2001, Karen Oberhauser of the University of Minnesota and collaborators surveyed four regions of the monarch's summer breeding range and showed that many more monarchs were produced in corn fields (on weedy milkweeds) compared with nonagricultural fields. Interestingly, although monarch "production" (eggs per area of field) was higher in nonagricultural areas, there is so much more land area devoted to corn fields than to natural areas, especially in Minnesota and Wisconsin, that overall it was estimated that many more monarchs were produced in corn fields. Although scaling up from field surveys to entire regions requires assumptions (for example, the amount of cropland, fields, and other potential habitat, let alone the overall abundance and density of milkweed), it is undeniable that croplands have been hugely important for monarchs, in part because milkweeds have often been abundant in crop fields or at their edges, and because croplands make up a very sizable fraction of our landscape across the entire continent.

Then, amid this demonstration of the importance of agricultural fields for monarchs and the rumblings about increased herbicide use to kill weeds, the milkweed limitation hypothesis gained momentum. In 2012, John Pleasants of the University of Iowa and Karen Oberhauser showed that across an eleven-year period between 1999 and 2010, not only did milkweed decline in agricultural areas (based on surveys in Iowa), but so too have monarch egg densities declined across the Midwest (based on data collected by citizen scientists), and the overwintering populations in Mexico also declined—all coincident with the adoption of herbicide-tolerant crops. Their argument was based on data, some assumptions and extrapolations, and the reasonable expectation that declining milkweed and declining monarchs are directly linked.

Yet, the causal link between adoption of herbicides and the decline of monarchs was still tenuous in my mind. It all made logical sense. And yet, it is a gnarly set of correlations that appear tangled. If monarchs and milkweeds are declining, and herbicide use is increasing, are there alternative explanations, or have we found the cause? Here is where the "pin-headed scientist" in me who wants more data clashes with the "concerned environmentalist" in me who panics as a crisis is brewing. I sometimes feel like both of these caricatures—wanting to be rigorous in the basis of our conclusions, but not wanting to wait to act until disaster occurs. Nonetheless, I typically try to play the part of the "ecologist," concerned and cautious, relying on scientific data and rigorous analyses, but trying to be realistic about what data can be collected.

SPURIOUS CORRELATIONS?

Then something happened that would change me forever. In the summer of 2014, I gave a lecture at the Rocky Mountain Biological Laboratory in Gothic, Colorado, and I also had agreed to give an evening public lecture in the nearby town of Crested Butte. After the lecture, an audience member asked about the decline of monarchs and the causes. Much to my surprise, he claimed that cellular telephone towers were causing tremendous disruptions in electromagnetic radiation, and that such disruptions were known to affect other long-distance migrants such as birds. To be honest, although I knew of some birds using electromagnetic fields as a map (see chapter 8), the comment seemed a bit on the quackery side of things. And to be clear, neither then, nor now, do I believe that cell phone usage is the cause of monarch declines. But this comment set me off on a mission to untangle the gnarly correlations described above.

The problem with monarch population data is that we are stuck with the messy information we have. We cannot do a clever experiment to figure out the cause of the monarchs' reduced numbers, and we cannot travel back in time to take better data. The scale of the problem stretches over decades and millions of square miles. The first thing I did was try to recreate the set of correlations between the proposed mechanisms of monarch declines and their numbers in Mexico. I tracked the adoption of herbicide-tolerant corn and soy-

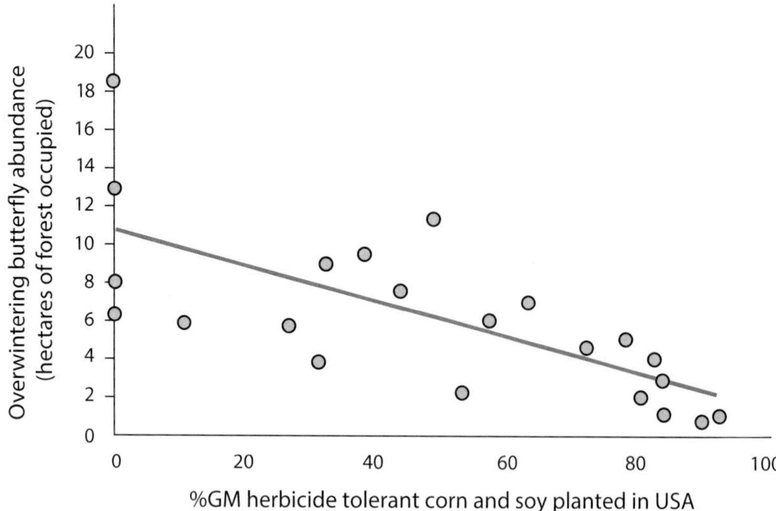

a

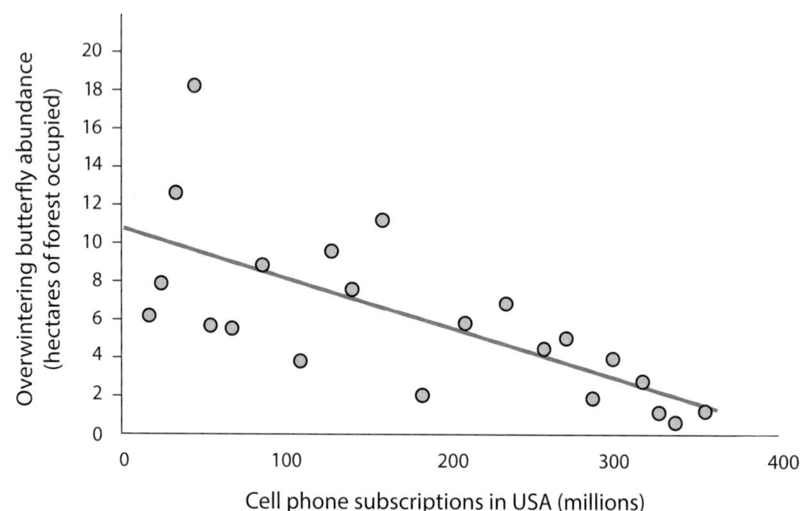

b

FIGURE 9.8. (a) Correlations between the percentage of genetically modified, herbicide-tolerant corn and soybeans planted each year in the United States and the monarch overwintering population size in Mexico; and (b) the number of cell phone subscriptions in the United States and the same estimate of monarch populations. Each point represents a year (1993–2014). Note that the points are not in chronological order, although there is a temporal decline in monarch numbers (see fig. 9.1). These correlations are indistinguishable, challenging our notion of what is causing the monarch decline.

bean crops and simply plotted these against monarch populations. The relationship is statistically significant and impressive (fig. 9.8a). In all, nearly 50 percent of the variation in monarch overwintering numbers can be explained by the adoption of herbicide-tolerant corn and soybeans. Note that in this figure's graphs, although the points are not in chronological order, the first few points, representing near zero percentages of herbicide-tolerant crops, are from the years 1993–96, and the last three points, representing the lowest monarch populations, are from the period 2012–14.

When I next plotted the number of cell phone subscriptions in the United States over the same time period, much to my surprise, the graph looked very similar (fig. 9.8b). In this case, 47 percent of the variation was explained. Although I do not believe that cell phones are the cause of monarch declines, this result gave me pause. When I tried correlating the success of the stock market (January values for the S&P 500) over the same period for predicting monarch numbers, I was again surprised: a significant correlation, although here explaining only 32 percent of the variation in monarch numbers, was found.

What could possibly be going on? What I have concluded is that any factor, whether it is the stock market value or the number of Twitter posts, that shows a directional year-to-year increase will show a statistically significant relationship with monarch overwintering numbers (because they have been declining in a directional way over the same years). Of course the decline of milkweed owing to genetically modified herbicide-tolerant crops is a more parsimonious explanation for monarch declines, but I felt that we must look more critically at the data.

PLANTING MILKWEED WON'T HALT THE MONARCH DECLINE

Working with Hidetoshi Inamine as part of his doctoral research at Cornell, we spent more than a year working through the existing data to try and piece together what is happening with monarchs. As discussed above, our conclusion was that the monarch decline is unlikely to be caused by lack of milkweed.

Several different approaches led us to this conclusion, the most important of which is that the steepness of the monarch decline begins in Mexico and becomes progressively weaker over the subsequent three to four generations. Each generation during the annual cycles depends on the previous generation, but each new generation appears to compensate for the decline in Mexico by increasing in abundance. All indications are that the decline itself is due to a shortfall that happens during the autumn migration or soon after (see also fig. 9.7, above). Based on the data, the problem occurs after monarchs are reared on milkweed. Here is what we know.

Several things are critical for monarchs during the autumn migration, including minimizing energy expenditure, accurate navigation to Mexico, and sustained habitats with nectar and resting spots. Additionally, the quality of the overwintering grounds themselves could impact the ultimate settling of monarchs for the winter. One recent observation is that the pace of the autumn migration has increased over the past decade, meaning the butterflies are covering more ground, and faster, although the cause of this effect is not known. Also, the sex ratio of monarchs at the overwintering sites has precipitously changed over the past three decades, from 53 percent female to 43 percent female. It is conceivable that sex differences in mass, flying ability, disease resistance, and lipid storage contribute to differences in survival and thus the shift in sex ratio.

Nectar is extremely important as the sole fuel source for adult monarchs, and nectar during the autumn migration is converted to all-important lipids to survive the winter. In the autumn, milkweeds are typically done flowering, so they are no longer an important resource for the adult migrants. Many plants in the daisy family (Asteraceae), however, especially the goldenrods and asters, are likely to be extremely important nectar sources in the autumn. Although it is difficult to quantify the impact of habitat fragmentation, development, and agricultural practices (including the use of herbicides and insecticides), there is no doubt that these factors have intensified, and that nectar sources may in fact be declining.

Of particular importance is what is happening during the autumn in the southern part of the migration, beginning in Texas. The least-well-understood part of the southern migration is what happens across the border in Mexico. Essentially nothing is known about the last 25–30 percent of the journey, covering seven hundred miles (one thousand kilometers). This part of the journey is important for many reasons, including the buildup of lipids, which are stored by converting floral nectar to fat. Yet, beginning in the autumn of 2010, Texas and areas of northern Mexico experienced the worst drought in at least sixty years, and it did not let up until the end of 2014. Lincoln Brower and colleagues have argued that this led to stress caused by the lack of nectar plants and limited the accumulation of lipids needed for overwintering. Attention to habitat quality and monarch health in this last third of the southern migration are especially necessary if we are to fully understand the decline, which I believe is occurring as a consequence of events that happen *after* summer breeding on milkweed in the north.

THE ECOLOGIST'S TREASURE TROVE OF INTERACTIVE EXPLANATIONS

A curse and blessing of being a conservation ecologist is the diversity of causes that impact life. Not only are life-forms incredibly diverse, but so too are the factors that may affect any one species. As we have seen above, habitat quality, food limitation (milkweed and floral nectar), and climate have each been implicated in the fluctuations and decline of monarch populations—not to mention insecticides and other factors that will be mentioned below. What we often believe to be going on, however, is that these factors might interact in ways that buffer or agitate population dynamics. Although winter storms can slash monarch populations by more than 70 percent, good spring weather can allow the population to recover completely, as occurred in 2001–2. Conversely, few nectar resources during the autumn migration, combined with poor forest quality in Mexico and reduced milkweed densities in the Midwest, could conspire to pummel monarch populations. Ecologists are challenged with not only parsing out the effects of individual factors, but trying to understand their combinations.

Other hypotheses have been put forward for negative environmental effects on monarchs. Insecticides, including the so-called neonicotinoids, have been suggested as reasons for the monarchs' decrease, although rigorous data to test this hypothesis are just now being collected. Invasive non-*Asclepias* plants, especially closely related species in the milkweed family (Apocynaceae), have been described as ecological "traps" that encourage monarchs to lay eggs but do not provide a suitable food for caterpillars. Increased incidence of monarch diseases and parasitoids (some of which were introduced to kill gypsy moths) can devastate monarch populations. The impact of ecotourism at the overwintering grounds has been a constant debate: although it can improve the livelihoods of local communities and thus strengthen restrictions on logging, it can also deteriorate the habitat so critical for monarch survival. In a typical year, nearly one hundred thousand visitors come to see the monarch overwintering sites. Even collisions with cars have been discussed. In the state of Illinois alone, perhaps two million monarchs are killed by cars during the southern migration each year. If we scaled this up across the states of their southern flight, the number could easily be as high as 20 million deaths a year! And finally, a role for electromagnetic disruptions, while not proven, has also been suggested as a possible contributor to monarch declines.

It's a wonder that any monarch, weighing about as much as a dollar bill, can make the journey, over many weeks, across so many hazards, to such a far-off and isolated destination. They are declining, especially in the number that reach and overwinter in that destination. But, unfortunately, the main cause, or the confluence of causes remains debatable. And if combinations of factors are responsible for the decline of monarchs, it may ultimately be quite difficult to parse out the specific causes.

WHITHER THE MONARCH?

In terms of public exposure, the monarchs had banner years in 2014 and 2015. A media blitz was catalyzed by several years of low monarch numbers. "Plant milkweed" was on the tip of a lot of tongues, and the US Fish and Wildlife Service was petitioned to list monarchs (and their migratory phenomenon) as threatened under the Endangered Species Act. The intimate relationship between monarchs and milkweeds has certainly been a theme of this book, and it is also a major theme in the current call for saving the monarch. It is perhaps ironic that I personally do not believe that milkweed is the limiting factor. Monarchs and milkweeds have coevolved, and monarchs are certainly dependent on milkweed. But the data on their population decline do not support the milkweed limitation hypothesis. Others may agree or disagree, but only time will tell.

As for the future of monarchs, new self-sustaining populations have sprung up in Spain, Hawaii, New Zealand, Australia, and likely other parts of the world. Some are migratory while others are not. They all depend on milkweed, which has been introduced to these new localities, mostly the tropical milkweed, *Asclepias curassavica*. As a result, the monarch as a species is not going extinct. Monarchs in North America are in trouble, but they will evade extinction as well. Monarchs will continue to sequester toxins and show off their aposematic colors. And they will continue to coevolve with milkweeds in an arms race. There is no question that many people around the world will continue to enjoy monarchs and milkweeds. Within their ecology lies a seemingly limitless trove of mysteries to be solved and learned from. As for their migration in North America, it is hard to argue with the fact that something isn't right. The onus is on us to understand the problem and to try and fix it.

ACKNOWLEDGMENTS

I have an ecologically inclined family, and they have been a source of support and inspiration for my love of the natural world. I am grateful to my parents for infusing me with a love for plants and for forcing me to spend time outdoors. My partner, Jennifer Thaler, another lover of insects on plants, has helped make this writing project fun. She provided an ear and inspiring discussion—and often a voice of reason. Our greatest joint project, that of raising two creatures, Jasper and Anna, not so small anymore, is one of the great pleasures of life, and I dedicate this book to the two of you. The award for the most hours spent listening to me yack, hash over ideas, and tell stories about monarchs and milkweeds goes to my in-laws, Pat Dubin and Ed Wurtz, who seem to have an insatiable desire to learn and a gleeful willingness to discuss anything and everything about monarchs and milkweed.

I next want to mention a few key people who have been instrumental in the development of my knowledge and thinking about monarchs and milkweeds. Without Jimmy Fordyce, and his initial encouragement, I might never have gotten onto this obsession. Early on, Steve Malcolm showed extreme generosity with ideas, technical expertise, plant collections, and general inspiration. Mark Fishbein has been a decade-long collaborator on many aspects of milkweed ecology and evolution, and it has been a great pleasure. Karen Oberhauser was remarkably generous with her time, knowledge, images, references, advice, and overall positive encouragement. Finally, I acknowledge Lincoln Brower, who has both been supportive of this project and provided incisive feedback on all of it. Even in disagreement, Lincoln remained objective and open.

The accumulation of knowledge, which in the modern academic system largely takes place at universities, is one of the great privileges and gifts bestowed by our society. This accumulation of knowledge typically does not happen through instruction per se, but through collaborative research, which especially in the natural sciences, involves undergraduate students, graduate students, postdoctoral researchers, and technical staff. I have been fortunate to work with a truly talented set of such scholars, and they are particularly noteworthy for how they have challenged me and for their contributions to the science presented in this book. In particular, the following have worked on monarchs and milkweeds, or otherwise taught me great biology through shared interactions in the laboratory and field: Jared Ali, Lina Arcila Hernandez, Rowan Barrett, Robin Bingham, Susan Cook-Patton, Graham Cox, Gaylord Desurmont, Collin Edwards, Jacob Elias, Alexis Erwin, Daniel Fines, William Godsoe, Jessica Goldstein, Natalie Griffiths, Karin Gustafsson, Rayko Halitschke, Amy Hastings, Katie Holmes, Daisy Johnson, Marc Johnson, Patty Jones, Emily Kearney, Anna Knight, Nile Kurashige, Marc Lajeunesse, Sophie Mao, Andrew McDowell, Rosanna McGuire, Kailen Mooney, John Parker, Eamonn Patrick, Georg Petschenka, Lisa Plane, Trey Ramsey, Sergio Rasmann, Alex Smith, Mike Stastny, Scott Taylor, Andrew Tuccillo, Pete Van Zandt, Marjorie Weber, Ellen Woods, and Tobias Züst.

Special thanks are due to Amy Hastings, the research support specialist in our lab for more than eight years, and without whom our research program would be far less rich. Finally, our local academic environment shapes our intellectual curiosity and the satisfaction with which we pursue our scholarship. Cornell University has been an especially inspiring place to be, and this is largely due to colleagues who have created a stimulating campus. I am grateful to be part of the "group of six" (Cornell faculty) who have made studying insects on plants so rewarding: Georg Jander, Andre Kessler, Katja Poveda, Rob Raguso, and Jennifer Thaler.

This book project was stimulated over the years by conversations with Tom Seeley and Harry Greene. I am also especially grateful to those that have helped me partially fill in some of my critical knowledge gaps about monarchs and milkweeds. Eduardo Rendón-Salinas and Karina Boege facilitated a visit to the monarch overwintering grounds in 2011, which was a magical and mystical experience. I thank Steve Wolf, Bruce Lewenstein, and Karin Gustafsson, who gave me a crash course in social science during a wonderful collaboration on the environmental sociology and public understanding of science surrounding monarchs. I have been fortunate to collaborate with Hidetoshi Inamine and Steve Ellner on aspects of monarch population dynamics, which has pushed the envelope of our knowledge about conserving these butterflies. Susanne Dobler introduced me to the molecular and physiological world of monarchs and other milkweed insects. And finally, my neighbor Matt Klemm fueled my naive interest in the history of science.

Several people read and commented on all of the chapters in this book, and I especially thank them: Lincoln Brower, Don Davis, Pat Dubin, Beth Gianfagna, Patty Jones, Steve Malcolm, Michael Smith, Jennifer Thaler, John Thompson, Marjorie Weber, and Ed Wurtz. My editor at Princeton University Press, Alison Kalett, provided insightful advice from the beginning to the end of the project. And my extended lab group read and commented on chapters, with especially helpful comments on parts of the book from Katie Holmes, Hidetoshi Inamine, John Maron, Renee Petipas, and Georg Petschenka.

For answering queries, discussion, reading individual chapters, or general support, I thank: Sonia Altizer, Betty and Homero Aridjis, Kentaro Arikawa, Kristen Baum, Jeff Belth, May Berenbaum, Michael Boppré, Deane Bowers, Steve Broyles, John Burns, Brenda Casper, Bryan Danforth, Jaap de Roode, Rodolfo Dirzo, Jason Dombroski, Gary Duncan, Paul Ehrlich, Jim Ellis, Richard Ellis, Dan Fagin, Tyler Flockhart, Jeff Glassberg, Patrick Guerra, Meena Haribal, Drew Harvell, Chris Ivey, Isaac Jamieson, Dan Janzen, Brian Johnston, Susan Kephart, Skip Kiphart, Dwight Kuhn, Martin LaBar, Fidencio Marbella, Dale McClung, John Pleasants, Robert Pyle, Peter Raven, Eduardo Rendón-Salinas, Steven Reppert, Jim Reveal, Leslie Ries, Georgios Scheiner-Bobis, William Schoner, Ellen Sharp, Suzanne Shaw, Dan Slayback, Jim Springer, Chip Taylor, Dru Thomas, Alex Travis, Dick Vane-Wright, David Wagner, Ingrid Waldron, Len Wassenaar, Kent Weakley, Roxanne Young, and Meron Zalucki. I am also grateful to the Cornell University library staff, who not only answered questions and helped with research, but occasionally unearthed re-

Ellen Woods, on assignment, photographing milkweed leaf hairs.

Illustrator Frances Fawcett, with ornithologist Scott Taylor, examining bird specimens for this project.

markably obscure literature unexpectedly. Betsy Blumenthal provided behind-the-scenes help at Princeton University Press.

The images and illustrations in the book come from many sources, acknowledged in the credits. However, special thanks are due to Ellen Woods and Frances Fawcett for providing the lion's share of photographs and illustrations, respectively. I have lived vicariously through Ellen's camera lens. When others have called me a camera bluff, she came to my rescue. And Frances has helped me realize many visual interpretations of the entire story. I have been very fortunate to work with such talented and monarch-and-milkweed-loving artists. Meena Haribal, Jay Hart, Georg Petschenka, and Tobias Züst also prepared specific figures for the project.

My research on monarchs and milkweeds has been supported by the following organizations (in chronological order): University of Toronto, Natural Sciences and Engineering Council of Canada, Canadian Foundation for Innovation, National Science Foundation (NSF), Cornell University, United States Department of Agriculture, and John Templeton Foundation. Saran Twombly of the NSF deserves special mention as somebody who looked out for me over the years.

The Atkinson Center for a Sustainable Future (ACSF) at Cornell University provided specific funding for the enhancement of this book. The leadership of ACSF, Frank DiSalvo, Graham Kerslick, and Alex Travis, were particularly helpful. Finally, the staff of my home department, Ecology and Evolutionary Biology, at Cornell University supported this project in many direct and indirect ways, and I am especially grateful to Carol Damm, Luanne Kenjerska, and Jennifer Robinson.

NOTES

CHAPTER 1: WELCOME TO THE MONARCHY

Page 1: Epigraph. Aridjis, Homero. 2002. "To a Monarch Butterfly," trans. George McWhirter, in George McWhirter and Betty Ferber, eds. *Eyes to See Otherwise*. New York: New Directions. Aridjis is a Mexican writer and environmental activist. He grew up near the monarchs' overwintering grounds and as a child would hike up a hillside behind his village (Contepec, near site #3 in fig. 8.10) to watch the butterflies, long before scientists had discovered their location. This poem was first published in 1971 as part of his book *El poeta niño*.

Page 3: The term "arms race" was first used H. B. Cott in his classic 1940 book, *Adaptive Coloration in Animals*. London: Methuen. He wrote: "Indeed, the primeval struggle of the jungle, and the refinements of civilized warfare, have here very much the same story to tell. In both realms we see the results of an armament race and an invention race, which has led to a state of preparedness for offence and defence as complex as it is interesting." Although such dynamics had been foreshadowed much earlier (for example, in Stahl, Ernst. 1888. *Pflanzen und Schnecken: Eine biologische Studie über die Schutzmittel der Pflanzen gegen Schneckenfraß* [Plants and snails: A biological study regarding protection of plants against slugs]. Jena: G. Fischer), the arms race analogy was cemented in a 1971 paper by two Cornell University ecologists, Robert Whittaker and Paul Feeny, that defined the emerging field of chemical ecology (Allelochemics: Chemical interactions between species, *Science* 171:757–70). In this book, I primarily discuss arms race coevolution. However, another form of coevolution, termed gene-for-gene (or allele matching), is also common in nature and has been summarized in Thompson, J. N., and J. J. Burdon. 1992. Gene-for-gene coevolution between plants and parasites. *Nature* 360:121–25.

Page 6: Figure 1.3. The biogeographic origins and distribution of monarch butterflies was recently elucidated in Zhan, S., et al. 2014. The genetics of monarch butterfly migration and warning colouration. *Nature* 514:317–21.

Page 7: In 1758, the great Swedish biologist Carolus Linnaeus, famous for creating a classification system for plants and animals, originally placed monarchs in the genus *Papilio* (i.e., the swallowtails). In older literature, monarchs were also referred to with other Latin names, including *Danaus archippus* (Fabricius, 1793), *Danaus menippe* (Hübner, 1816), and *Anosia plexippus* (Dyar, 1903).

Page 12: The proportion of plant production that is consumed annually by herbivores has been estimated to be between 5 and 20 percent. Most quantitative estimates focus on leaf consumption and arrive at values between 5 and 10 percent. I have argued that the number is likely closer to 20 percent because of the many hidden consumers of plants: the aphids and other insects that suck the sap, and various animals that eat roots.

Page 12: Figure 1.6. This image was first developed in Wheeler, Q. D. 1990. Insect diversity and cladistic constraints. *Annals of the Entomological Society of America* 83:1031–47. Note that only macroscopic organisms are depicted. We currently lack a strong understanding of microbial diversity, although it surely is much greater than that of macroscopic organisms. For recent information on estimates of species diversity on Earth, see Mora, C., et al. 2011. How many species are there on Earth and in the ocean? *PLoS Biology* 9:e1001127, http://dx.doi.org/10.1371/journal.pbio.1001127.

Page 13: The classic paper on "strong inference" is Platt, J. R. 1964. Strong inference. *Science* 146:347–53. The tongue-in-cheek reference to correlations and Nobel prizes is Messerli, F. H. 2012. Chocolate consumption, cognitive function, and Nobel laureates. *New England Journal of Medicine* 367:1562–64.

Page 15: Jimmy Fordyce's master's thesis was one of the first papers published on milkweed's stem-feeding weevil: Fordyce, J. A., and S. B. Malcolm. 2000. Specialist weevil, *Rhyssomatus lineaticollis*, does not spatially avoid cardenolide defenses of common milkweed by ovipositing into pith tissue. *Journal of Chemical Ecology* 26:2857–74.

Pages 15–16: Between 2000 and 2013, the Union of Concerned Scientists used as its logo the monarch butterfly on top of a globe, but the organization recently changed its emblem, removing both images. In the words of one of its climate scientists, Doug Boucher, "The butterfly was a wonderful spokesinsect, but metamorphosis is something that happens to all of us."

Page 16: A recent study on the valuation of monarchs by the American public suggests a strong interest and willingness to take part in the future of this species: Diffendorfer, J. E., et al. 2014. National valuation of monarch butterflies indicates an untapped potential for incentive-based conservation. *Conservation Letters* 7:253–62, http://onlinelibrary.wiley.com/doi/10.1111/conl.12065/full.

Page 17: For Emerson's definition of a weed, see Emerson, R.W. 2010. *The Later Lectures of Ralph Waldo Emerson, 1843–1871*, vol. 2, ed. R. A. Bosco and J. Myerson. Athens: University of Georgia Press.

Page 19: Figure 1.8. This figure and phylogeny was modified from Petschenka, G., et al. 2013. Stepwise evolution of resistance to toxic cardenolides via genetic substitutions in the Na^+/K^+-ATPase of milkweed butterflies (Lepidoptera: Danaini). *Evolution* 67:2753–61. Although an active area of research, the latest published information on the evolutionary relationships between the milkweed butterflies is available in two publications: Smith, D.A.S., G. Lushai, and J.A.A. Allen. 2005. A classification of *Danaus* butterflies (Lepidoptera: Nymphalidae) based upon data from morphology and DNA. *Zoological Journal of the Linnean Society* 144:191–212; and Brower, A. V., et al. 2010. Phylogenetic relationships among genera of Danaine butterflies (Lepidoptera: Nymphalidae) as implied by morphology and DNA sequences. *Systematics and Biodiver-*

sity 8:75–89. Most other Danaini are far less studied than monarchs. Nonetheless, a recent scholarly work chronicles the biology of the "African monarch" (also known as the plain tiger, *Danaus chrysippus*), and this book is especially interesting, as it compares the biology of several closely related *Danaus* species: Smith, D. A. 2014. *African Queens and Their Kin: A Darwinian Odyssey*. Taunton, UK: Brambleby Books.

Page 20: Figure 1.9. The systematic investigations of *Asclepias* have been led by Mark Fishbein, and this figure was developed based on what is presented in two publications: Fishbein, M., et al. 2011. Phylogenetic relationships of *Asclepias* (Apocynaceae) inferred from non-coding chloroplast DNA sequences. *Systematic Botany* 36:1008–23; and Agrawal, A. A., et al. 2014. Macroevolutionary trends in the defense of milkweeds against monarchs: Latex, cardenolides, and tolerance of herbivory, in K. Oberhauser, K. Nail, and S. Altizer, eds. *Monarchs in a Changing World: Biology and Conservation of an Iconic Insect*. Ithaca, NY: Cornell University Press, 47–59. The diversity of species included in the genus *Asclepias* has been a matter of debate. Under some classifications, only the approximately 130 North American species have been included. More recently, Fishbein has advocated the inclusion of the more than 250 species native to Africa (often included in the genus *Gomphocarpus*) in *Asclepias*. Although the boundary of which species to include in which genus is somewhat arbitrary, if the North American species are monophyletic (i.e., include *all* the descendants of a common ancestor), I favor separating the North American species into one genus (*Asclepias*).

Page 21: Monarchs are representative of other butterflies and moths in that their specialization on milkweed parallels the level of specialization seen in other caterpillars. An important study outlines the distribution of host plant specificity among diverse Lepidoptera across the world: Dyer, L. A., et al. 2007. Host specificity of Lepidoptera in tropical and temperate forests. *Nature* 448:696–99.

CHAPTER 2: THE ARMS RACE

Page 22: Epigraph. The first edition (1859) text of Darwin's *On the Origin of Species* is available at Project Gutenberg, http://www.gutenberg.org/files/1228/1228-h/1228-h.htm.

Page 23: The great evolutionary biologist Theodosius Dobzhansky famously argued in 1973 that nothing in biology makes sense except in the light of evolution in an essay published in *American Biology Teacher* 35:125–29.

Pages 25–26: The original study by Stahl was published as a book: Stahl, E. 1888. *Pflanzen und Schnecken: Eine biologische Studie über die Schutzmittel der Pflanzen gegen Schneckenfraß* [Plants and snails: A biological study regarding protection of plants against slugs]. Jena: G. Fischer. For other historical nuggets about coevolution, see the second note in chapter 1.

Pages 28–29: It is widely known among milkweed biologists that monarchs are not good pollinators of most milkweeds; this is evident after even brief observations of monarchs on milkweed flowers. Several published studies have addressed the issue of which species of flower visitors are the best pollinators: Fishbein, M., and D. L. Venable. 1996. Diversity and temporal change in the effective pollinators of *Asclepias tuberosa*. *Ecology* 77:1061–73; Kephart, S., and K. Theiss. 2004. Pollinator-mediated isolation in sympatric milkweeds (*Asclepias*): Do floral morphology and insect behavior influence species boundaries? *New Phytologist* 161:265–77, http://onlinelibrary.wiley.com/doi/10.1046/j.1469-8137.2003.00956.x/full; Willson, M. F., and Bertin, R. I. 1979. Flower-visitors, nectar production, and inflorescence size of *Asclepias syriaca*. *Canadian Journal of Botany* 57:1380–88; Jennersten, O., and D. H. Morse. 1991. The quality of pollination by diurnal and nocturnal insects visiting common milkweed, *Asclepias syriaca*. *American Midland Naturalist* 125:18–28; Howard, A. F. and E. M. Barrows. 2014. Self-pollination rate and floral-display size in *Asclepias syriaca* (common milkweed) with regard to floral-visitor taxa. *BMC Evolutionary Biology* 14:144, http://bmcevolbiol.biomedcentral.com/articles/10.1186/1471-2148-14-144.

Page 30: For how President Obama lumped honeybees and monarchs as critical pollinators, see the presidential memorandum at http://www.whitehouse.gov/the-press-office/2014/06/20/presidential-memorandum-creating-federal-strategy-promote-health-honey-b.

Page 32: The translation of Léo Errera's comments on alkaloids is in Hartmann, T. 2008. The lost origin of chemical ecology in the late 19th century. *Proceedings of the National Academy of Sciences* 105:4541–46, http://www.pnas.org/content/105/12/4541.full. This history of chemical ecology is an excellent and thorough account of the origins of this interdisciplinary science.

Page 32: The call to solve the "problem" of monarch toxicity is in Poulton, E. B. 1914. Mimicry in North American butterflies: A reply. *Proceedings of the Academy of Natural Sciences of Philadelphia* 66:161–95, http://www.jstor.org/stable/pdf/4063560.pdf.

Pages 32–33: A brief biography of Miriam Rothschild may be found in van Emden, H. F., and J. Gurdon. 2006. Dame Miriam Louisa Rothschild CBE. *Biographical Memoirs of Fellows of the Royal Society* 52:315–30, http://rsbm.royalsocietypublishing.org/content/52/315.full.pdf. Rothschild recounted her initial correspondence with Tadeus Reichstein (1897–1996) in her 1999 brief biography of the Nobel laureate in *Biographical Memoirs of Fellows of the Royal Society* 45:451–67. The original description of cardenolides in plants was in Reichstein, T. 1951. Chemie der herzaktiven glykoside [Chemistry of cardioactive glycosides]. *Angewandte Chemie* 63:412–21. Their first collaboration was published as an abstract: Rothschild, M., T. Reichstein, et al. 1966. Poisons in aposematic insects. *Royal Society Conversazione* 12:10. There is no record of what was shown in the research presentation by Rothschild, and as such it does not constitute a peer-reviewed publication that would be given precedence in terms of the order of discoveries. The "delightfully disheveled garden" is quoted from my personal correspondence with Steven Malcolm.

Pages 33–37: The four keystone papers outlined in this section are (1) Parsons, J. A. 1965. A digitalis-like toxin in the monarch butterfly, *Danaus plexippus* L. *Journal of Physiology* 178:290–304; (2) Brower, L. P., J. van Brower, and J. M. Corvino. 1967. Plant poisons in a terrestrial food chain. *Proceedings of the National Academy of Sciences USA* 57:893–98, http://www.pnas.org/content/57/4/893.short; (3) Reichstein, T. 1967. Cardenolide (herzwirksame glykoside) als abwehrstoffe bei insekten [Cardenolides (cardioactive glycosides) as defensive substances in insects]. *Naturwissenschaftliche Rundschau* 20:499–511 (Note: all coauthors, including Rothschild, Parsons, and Brower, were apparently removed from the author list by the editor of the journal because this was a transcript of a lecture; instead these authors and their contributions were listed above the typical acknowledgments); and (4) Reichstein, T., J. Von Euw, J. A. Parsons, and M. Rothschild. 1968. Heart poisons in the monarch butterfly. *Science* 161:861–66. In this latter paper, reference no. 24 lists all of the authors of the 1967 paper published in *Naturwissenschaftliche Rundschau*. Additionally, in personal correspondence with me, Lincoln Brower stated that he withdrew his name from the Reichstein et al. 1968 *Science* paper. It is cited as "in preparation" in Brower et al. (1967) and includes Brower as an author at that earlier stage.

Page 38: In this quote from *On the Origin of Species*, Darwin mentions the German natural philosopher Johann Wolfgang von Goethe (1749–1832), one of the greatest writers and scholars in history, whose theories also foreshadowed the concept of biological evolution. Concerning trade-offs, Darwin recognized a distinction between a trade-off caused by energy diversion versus a negative effect that was favored by natural selection. Indeed, the connection of any two arrows may be driven by the fact that they share a particular energy stream or because natural selection favored the connection: when one trait is used, the other is not needed. Technical details of this distinction are reviewed in Agrawal, A. A., J. K. Conner, and S. Rasmann. 2010. Tradeoffs and adaptive negative correlations in evolutionary ecology, in M. Bell, et al., eds. *Evolution after Darwin: The First 150 Years*. Sunderland, MA: Sinauer Associates, 243–68.

Page 39: Figure 2.6. This model of resource allocation was a step up from a typically accepted binary Y-tube of allocation trade-offs between two traits. The multiple arrows model was developed and presented in Karban, R., and I. T. Baldwin. 1997. *Induced Responses to Herbivory*. Chicago: University of Chicago Press.

Page 40: Figure 2.7. Costs of producing cardenolides (in terms of a trade-off with plant growth) were first reported by my laboratory: Züst, T., S. Rasmann, and A. A. Agrawal. 2015. Growth-defense tradeoffs for two major anti-herbivore traits of the common milkweed *Asclepias syriaca*. *Oikos* 124:1404–15.

Page 40: The cricket story is outlined in Zuk, M., L. W. Simmons, and L. Cupp. 1993. Calling characteristics of parasitized and unparasitized populations of the field cricket *Teleogryllus oceanicus*. *Behavioral Ecology and Sociobiology* 33:339–43.

CHAPTER 3: THE CHEMISTRY OF MEDICINE AND POISON

Page 43: Epigraph. In his classic book, *On the Nature of Things* (50 BCE), the Epicurean philosopher Lucretius gives a further example of one animal's food being another's poison:

> Again, fierce poison is the hellebore
> To us, but puts the fat on goats and quails.

In this case, we know that hellebores are toxic to humans owing to their cardiac glycosides (bufadienolides in particular), but it is unclear if goats and quails are truly not affected.

Pages 43–44: Millspaugh, Charles. 1892. *Medicinal Plants*. Philadelphia: J. C. Yorston and Co., 135-3.

Pages 45–46: For the hypothesis about how cardenolides affected Van Gogh's vision, see Lee, T. C. 1981. Digitalis intoxication? *Journal of the American Medical Association* 245: 727–29.

Pages 46–47: Historical accounts of the use of cardenolides are provided in Hoffman, B. F., and J. T. Bigger Jr. 1980. Cardiovascular drugs: Digitalis and allied cardiac glycosides, in A. G. Gilman, L. S. Goodman, and A. Gilman, eds. *The Pharmacological Basis of Therapeutics*. New York: Macmillan, 729–60.

Pages 46–47: Withering, William. 1785. *An Account of the Foxglove and Some of Its Medical Uses*. Birmingham: M. Swinney, 192, available at Project Gutenberg, http://www.gutenberg.org/files/24886/24886-h/24886-h.htm.

Page 48: For current information on cardenolides as anticancer therapeutics, see Prassas, I., and E. P. Diamandis. 2008. Novel therapeutic applications of cardiac glycosides. *Nature Reviews Drug Discovery* 7:926–35. Many more such publications appear monthly in medical journals.

Pages 49–52: For a scientific review of cardenolide chemistry and its ecological effects, see Agrawal, A. A., et al. 2012. Toxic cardenolides: Chemical ecology and coevolution of specialized plant-herbivore interactions. *New Phytologist*, 194:28–45, http://onlinelibrary.wiley.com/doi/10.1111/j.1469-8137.2011.04049.x/full.

Page 51: The discovery of the sodium pump was made in 1957 by the Danish chemist Jens Christian Skou, and he was awarded the Nobel Prize in Chemistry forty years later. Thus original study is Skou, J. C. 1957. The influence of some cations on an adenosine triphosphatase from peripheral nerves. *Biochimica et Biophysica Acta* 23:394–401.

Page 53: The discovery of ouabain as a human hormone was reported in Hamlyn, J. M., et al. 1991. Identification and characterization of a ouabain-like compound from human plasma. *Proceedings of the National Academy of Sciences* 88:6259–63, http://www.pnas.org/content/88/14/6259.full.pdf. How cardenolides affect the human heart is summarized in Schoner, W., and G. Scheiner-Bobis. 2007. Endogenous and exogenous cardiac glycosides and their mechanisms of action. *American Journal of Cardiovascular Drugs* 7:173–89.

Page 56: Digitoxin poisoning is reported in Lely, A. H., and C.H.J. Van Enter. 1970. Large-scale digitoxin intoxication. *British Medical Journal* 3:737–40, http://www.bmj.com/content/bmj/3/5725/737.full.pdf. And, although many cases of cardenolide poisoning exist, two plants in the milkweed family have been particularly well studied, and their toxic effects are reported in Saravanapavananthan, N., and J. Ganeshamoorthy. 1988. Yellow oleander poisoning—a study of 170 cases. *Forensic Science International* 36:247–50; and Gaillard, Y., A. Krishnamoorthy, and F. Bevalot. 2004. *Cerbera odollam*: A "suicide tree" and cause of death in the state of Kerala, India. *Journal of Ethnopharmacology* 95:123–26.

Page 57: For cardenolides reported in floral nectar of milkweeds, see Manson, J. S., et al. 2012. Cardenolides in nectar may be more than a consequence of allocation to other plant parts: A phylogenetic study of *Asclepias*. *Functional Ecology*, 26:1100–1110, http://onlinelibrary.wiley.com/doi/10.1111/j.1365-2435.2012.02039.x/full.

Page 58: For direct effects of cardenolides on caterpillars, see Karowe, D. N., and V. Golston. 2006. Effect of the cardenolide digitoxin on performance of gypsy moth (*Lymantria dispar*) (Lepidoptera: Lymantriidae) caterpillars. *Great Lakes Entomologist* 39:34. Also, Fukuyama, Y., et al. 1993. Insect growth inhibitory cardenolide glycosides from *Anodendron affine*. *Phytochemistry* 32:297–301. Injections of cardenolides have a negative impact on many herbivores, including some milkweed feeders (who normally would not absorb cardenolides into their bodies). For example, see Rafaeli-Bernstein, A.D.A., and W. Mordue. 1978. The transport of the cardiac glycoside ouabain by the Malpighian tubules of *Zonocerus variegatus*. *Physiological Entomology* 3:59–63. The image in figure 3.6 was derived from a recent study: Petschenka, G., and A. A. Agrawal. 2015. Milkweed butterfly resistance to plant toxins is linked to sequestration, not coping with a toxic diet. *Proceedings of the Royal Society B* 282:1865. Finally, even monarchs can be deterred by high levels of some cardenolides. In the following study, digitoxin (but not ouabain) was shown to have negative effects on monarchs' growth when ingested from compounds painted on leaves: Rasmann, S., M. D. Johnson, and A. A. Agrawal. 2009. Induced responses to herbivory and jasmonate in three milkweed species. *Journal of Chemical Ecology* 35:1326–34.

Pages 59–60: The discovery of how monarchs achieve their insensitivity to cardenolides was reported in Holzinger, F., and M. Wink. 1996. Mediation of cardiac glycoside insensitivity in the monarch butterfly (*Danaus plexippus*): Role of an amino acid substitution in the ouabain binding site of Na^+, K^+-ATPase. *Journal of Chemical Ecology* 22:1921–37.

CHAPTER 4: WAITING, MATING, AND MIGRATING

Page 63: Epigraph. Excerpt from "Stopping by Woods on a Snowy Evening" is from *The Poetry of Robert Frost*, ed. Edward Connery Lathem. Copyright © 1923, 1969 by Henry Holt and Company, copyright © 1951 by Robert Frost. Used by permission of Henry Holt and Company, LLC. All rights reserved.

Page 67: The work described on cold-induced change in migratory direction was published in Guerra, P. A., and S. M. Reppert. 2013. Coldness triggers northward flight in remigrant monarch butterflies. *Current Biology* 23:419–23, http://www.sciencedirect.com/science/article/pii/S0960982213000870.

Pages 67–71: Rothschild's choice commentary on monarch mating behavior was outlined in an obscure article: Rothschild, M. 1978. Hell's angels. *Antenna* 2:38–39. Coercive mating in monarchs was first outlined in detail by Pliske, T. E. 1975. Courtship behavior of the monarch butterfly, *Danaus plexippus* L. *Annals of the Entomological Society of America* 68:143–51. Building on Pilske's original observations, two articles have outlined the mating behavior and hypotheses to explain it: Oberhauser, K., and D. Frey. 1999. Coercive mating by overwintering male monarch butterflies, in W. A. Haber et al., eds. *1997 North American Conference on the Monarch Butterfly*. Montreal: Commission for Environmental Cooperation, 67–78; and Brower, L. P., et al. 2007. Monarch sex: Ancient rites or recent wrongs? *Antenna* 31:12–18.

Page 72: For a discussion of monarch lipid reserves and butterflies that do not migrate back to the United States in the spring, see Alonso-Mejía, A., et al. 1997. Use of lipid reserves by monarch butterflies overwintering in Mexico: Implications for conservation. *Ecological Applications* 7:934–47.

Page 73: For a classic entomology textbook and a clear statement of migratory hypotheses for monarchs, see Comstock, J. H., and A. B. Comstock. 1904. How to Know the Butterflies. New York: Appleton & Co. A scanned copy, from Cornell University is available at http://www.biodiversitylibrary.org/item/116129#page/12/mode/1up 5.

Pages 74–77: See Urquhart, F.A. 1952. Marked monarchs. *Natural History* 61:226–29. Although the article is credited to F. A. Urquhart only, he states in his 1960 book and again in his 1987 book that Nora Urquhart wrote the article. They clearly worked as a very effective team. I have tried to credit them both where possible, but her contributions are unclear in some cases. The first report on their citizen science studies is in Urquhart, F. A. 1955. Report on the studies of the movements of the monarch butterfly in North America. Publication of the Royal Ontario Museum, Toronto (fig. 4.6 is drawn from this report). Later Fred and Nora Urquhart would produce an annual newsletter, *Insect Migration Studies* (1964–94), available at http://www.monarchwatch.org/read/articles/index.htm. See also Urquhart, F.A. 1960. *The Monarch Butterfly*. Toronto: University of Toronto Press. A new book, which included a revision of the original text, but also details of new biology, including discovery of the overwintering grounds, was released twenty-seven years later: Urquhart, F. A. 1987. *The Monarch Butterfly: International Traveler*. Chicago: Nelson-Hall. Urquhart was often a contrarian, posing alternative hypotheses for the migration and other aspects of monarch biology. A well-known example is his disbelief in monarch toxicity and mimicry by viceroy butterflies, outlined in his 1960 book and also in the following report: Urquhart, F. A. 1957. A discussion of Batesian mimicry: As applied to the monarch and viceroy butterflies. Toronto: University of Toronto Press for the Division of Zoology and Palaeontology, Royal Ontario Museum.

Pages 78–79: Cardenolide fingerprinting was outlined in Roeske, C.N., et al. 1976. Milkweed cardenolides and their comparative processing by monarch butterflies (*Danaus plexippus* L.), in J. M. Wallace and R. L. Mansell, eds. *Biochemical Interactions between Plants and Insects*. Recent Advances in Phytochemistry, vol. 10. New York: Academic Press, 93–167. Later it was used in the important publication: Malcolm, S. B., B. J. Cockrell, and L. P. Brower. 1993. Spring recolonization of eastern North America by the monarch butterfly: Successive brood or single sweep migration? in S. B. Malcolm and M. P. Zalucki, eds. *Biology and Conservation of the Monarch Butterfly*. Science Series 38. Los Angeles: Natural History Museum of Los Angeles County, 253–67. This study reported that 92 percent of overwintering butterflies had the *Asclepias syriaca* cardenolide fingerprint. Nonetheless, other methods, including isotopic measures (see next note), suggest that this may be a slight overestimate, or at least that the percentage is variable year-to-year. Other methods for assessing the spring migration, including wing wear and modeling the monarch's temperature-dependent generation time were summarized in Cockrell, B. J., S. B. Malcolm, and L. P. Brower. 1993. Time, temperature, and latitudinal constraints on the annual recolonization of eastern North America by the monarch butterfly, in S. B. Malcolm and M. P. Zalucki, eds. *Biology and Conservation of the Monarch Butterfly*. Science Series 38. Los Angeles: Natural History Museum of Los Angeles County, 233–51.

Pages 79–81: The original paper outlining the isotopic method for tracking monarch movements is Wassenaar, L. I., and K. A. Hobson. 1998. Natal origins of migratory monarch butterflies at wintering colonies in Mexico: New isotopic evidence. *Proceedings of the National Academy of Sciences* 95:15,436–39, http://www.pnas.org/content/95/26/15436.full.pdf. And the most recent work, including the source for figure 4.7, is Flockhart, D. T., et al. 2013. Tracking multi-generational colonization of the breeding grounds by monarch butterflies in eastern North America. *Proceedings of the Royal Society B* 280:1087, http://rspb.royalsocietypublishing.org/content/280/1768/20131087. Other recent research suggests that as many as 10 percent of spring migrating monarchs may make it back to the northern parts of the breeding region; see Miller, N. G., et al. 2012. Migratory connectivity of the monarch butterfly (*Danaus plexippus*): Patterns of spring re-colonization in eastern North America. *PLoS One* 7:e31891, http://dx.doi.org/10.1371/journal.pone.0031891.

Pages 83–86: Important aspects of monarch oviposition behavior and its relation to plant chemistry were outlined in the PhD dissertation work of Meena Haribal, conducted at Cornell University under the guidance of Alan Renwick, and published in two important papers: Haribal, M., and J.A.A. Renwick. 1998. Differential postalightment oviposition behavior of monarch butterflies on *Asclepias* species. *Journal of Insect Behavior* 11:507–38; and Haribal, M., and J.A.A. Renwick. 1998. Identification and distribution of oviposition stimulants for monarch butterflies in hosts and nonhosts. *Journal of Chemical Ecology* 24:891–904. In addition, Kentaro Arikawa has described photoreceptors on the abdomen of several butterflies, including Danaines, which may be used to assess plant chemistry. See Arikawa, K. 2001. Hindsight of butterflies: The *Papilio* butterfly has light sensitivity in the genitalia, which appears to be crucial for reproductive behavior. *BioScience* 51:219–25.

Page 84: The quotation is from Eisner, T., and J. Meinwald. 1995. Chemical ecology. *Proceedings of the National Academy of Sciences* 92:14–18, http://www.pnas.org/content/92/1/14.full.pdf.

Pages 84–85: Many authors and naturalists alike have noted the preference of monarch butterflies for oviposition on resprouted plants. Despite some initial chemical analyses, the differences in oviposition stimulation have not been found; see Bergström, G., et al. 1994. Oviposition by butterflies on young leaves: Investigation of leaf volatiles. *Chemoecology* 5:147–158. For effects of fire on resprouting and oviposition, see Baum, K. A. and W. V. Sharber. 2012. Fire creates host plant patches for monarch butterflies. *Biology Letters* 8:968–71, http://rsbl.royalsocietypublishing.org/content/8/6/968.full. Finally, mowing has been suggested as a management strategy for improving milkweed quality for monarchs, but the actual impacts on larval growth and fitness, and ultimately on regional butterfly production have not been studied. See Fischer, S. J., et al. 2015. Enhancing monarch butterfly reproduction by mowing fields of common milkweed. *American Midland Naturalist* 173:229–40.

Page 85: The pattern of monarch egg-laying in North America was first outlined in detail in Borkin, S. 1982. Notes on shifting distributing patterns and survival of immature *Danaus plexippus* (Lepidoptera: Danaidae) on the food plant *Asclepias syriaca*. *Great Lakes Entomologist* 15:199–206. In the same year, a study reported a quite different pattern (many eggs per plant) in Australia, where monarchs were introduced: Zalucki, M. P., and R. L. Kitching. 1982. Dynamics of oviposition in *Danaus plexippus* (Insecta, Lepidoptera) on milkweed, *Asclepias* Spp. *Journal of Zoology* 198:103–16.

Page 87: The painstaking observational work of addressing what plants monarchs oviposit on in the field and assessing their chemical attributes (compared with the chemical attributes of plants not accepted for oviposition) was led by Myron Zalucki in several studies, two of which are Zalucki, M. P., L. P. Brower, and S. B. Malcolm. 1990. Oviposition by *Danaus plexippus* in relation to cardenolide content of three *Asclepias* species in the southeastern USA. *Ecological Entomology* 15:231–40; and Oyeyele, S. O., and M. P. Zalucki. 1990. Cardiac glycosides and oviposition by *Danaus plexippus* on *Asclepias fruticosa* in south-east Queensland (Australia), with notes on the effect of plant nitrogen content. *Ecological Entomology* 15:177–86.

Page 88: Jessamyn Manson was the first to describe cardenolides in milkweed nectar in a collaborative project with my laboratory: Manson, J. S., et al. 2012. Cardenolides in nectar may be more than a consequence of allocation to other plant parts: A phylogenetic study of *Asclepias*. *Functional Ecology* 26:1100–1110, http://onlinelibrary.wiley.com/doi/10.1111/j.1365-2435.2012.02039.x/full. Recent work from my laboratory on nectar cardenolides will appear in Jones, P. L., and A. A. Agrawal. 2016. Consequences of secondary compounds in nectar for mutualist bees and antagonist butterflies. *Ecology* (in press), doi: 10.1002/ecy.1483.

CHAPTER 5: HATCHING AND DEFENDING

Page 90: The original article posing the Red Queen's hypothesis is Van Valen, L. 1973. A new evolutionary law. *Evolutionary Theory* 1:1–30. Interestingly, Van Valen started the journal *Evolutionary Theory* himself; it was typed and photocopied in the administrative offices of his home department at the University of Chicago for decades. This classic study was in the first issue.

Pages 92–94: Our work on trichome and wax evolution on milkweed leaf surfaces was published in Agrawal, A. A., et al. 2009. Phylogenetic ecology of leaf surface traits in the milkweeds (*Asclepias* spp.): Chemistry, ecophysiology, and insect behaviour. *New Phytologist* 183:848–67, http://onlinelibrary.wiley.com/doi/10.1111/j.1469-8137.2009.02897.x/full.

Pages 95–99: I reviewed the history, biology, and chemistry of latex in Agrawal, A. A., and K. Konno. 2009. Latex: A model for understanding mechanisms, ecology, and evolution of plant defense against herbivory. *Annual Review of Ecology, Evolution and Systematics* 40:311–31. The early classic experiments were published in Kniep, H. 1905. Über die Bedeutung des Milchsafts der Pflanzen [On the importance of the milk sap of plants]. *Flora oder Allgemeine Botanische Zeitung* (Jena) 94:129–205. The best empirical summary of the effects of latex on monarchs comes from the massive set of experiments published in Zalucki, M. P., et al. 2001. It's the first bites that count: Survival of first-instar monarchs on milkweeds. *Austral Ecology* 26:547–55.

Page 96: Joseph James's comments on milkweed latex and defense were published in James, J. F. 1887. The milkweeds. *American Naturalist* 21:605–15.

Page 97: Figure 5.3. Results shown here were reported in Zalucki, M. P., and S. B. Malcolm. 1999. Plant latex and first-instar monarch larval growth and survival on three North American milkweed species. *Journal of Chemical Ecology* 25:1827–42. As might be expected, severing the latex-delivering canals in milkweed species with little latex (for example, butterfly weed, *Asclepias tuberosa*) had little impact on monarchs' growth.

Page 101: These remarkable observations on the function of circle trenches in other milkweed butterflies were published in DeVries, P. J. 1991. Foam barriers, a new defense against ants for milkweed butterfly caterpillars (Nymphalidae: Danainae). *Journal of Research on the Lepidoptera* 30:261–66.

Pages 102–103: The quotation on first instar monarch behavior and figure 5.6 are from Zalucki, M. P., and L. P. Brower. 1992. Survival of first instar larvae of *Danaus plexippus* (Lepidoptera: Danainae) in relation to cardiac glycoside and latex content of *Asclepias humistrata* (Asclepiadaceae). *Chemoecology* 3:81–93.

Pages 106–108: Jo Brewer's comments were published in her column Short lived phenomena. 1997. *News of the Lepidopterists' Society* 4:7. Miriam Rothschild had a different interpretation, published in the same journal and same year: The cat-like caterpillar. 1997. *News of the Lepidopterists' Society* 6:9, both at http://images.peabody.yale.edu/lepsoc/nls.

Pages 109–110: Evidence for negative effects of cardenolides on monarchs comes from four important publications: Seiber, J. N., et al. 1980. Pharmacodynamics of some individual milkweed cardenolides fed to larvae of the monarch butterfly (*Danaus plexippus* L.). *Journal of Chemical Ecology*, 6:321–39; Zalucki, M. P., L. P. Brower, and A. Alonso. 2001. Detrimental effects of latex and cardiac glycosides on survival and growth of first-instar monarch butterfly larvae *Danaus plexippus* feeding on the sandhill milkweed *Asclepias humistrata*. *Ecological Entomology* 26:212–24; Agrawal, A. A. 2005. Natural selection on common milkweed (*Asclepias syriaca*) by a community of specialized insect herbivores. *Evolutionary Ecology Research* 7:651–67; and Rasmann, S., M. D. Johnson, and A. A. Agrawal. 2009. Induced responses to herbivory and jasmonate in three milkweed species. *Journal of Chemical Ecology* 35:1326–34.

Pages 110–112: Rick Karban has written two outstanding books on plant responses to the environment. The first (Karban, R., and I. T. Baldwin. 1997. *Induced Responses to Herbivory*. Chicago: University of Chicago Press) is the classic treatise summarizing plant responses to herbivory and was in many respects the guiding document for my own PhD research. The second (Karban, R. 2015. *Plant Sensing and Communication*. Chicago: University of Chicago Press) includes aspects of plant responses to herbivory but much more broadly takes on the issue of plants as behavioral organisms.

Page 111: Figure 5.9. The data shown are from Agrawal, A. A., E. T. Patrick, and A. P. Hastings. 2014. Tests of the coupled expression of latex and cardenolide plant defense in common milkweed (*Asclepias syriaca*). *Ecosphere* 5:art126, http://www.esajournals.org/doi/pdf/10.1890/ES14-00161.1.

Pages 116–117: Alternative forms of defense in the milkweed include regrowth after defoliation (tolerance of herbivory by storage underground) and calling in predatory insects that eat monarchs. Three important publications on these topics are Hochwender, C. G., R. J. Marquis, and K. A. Stowe. 2000. The potential for and constraints on the evolution of compensatory ability in *Asclepias syriaca*. *Oecologia* 122:361–70; Agrawal, A. A., and M. Fishbein. 2008. Phylogenetic escalation and decline of plant defense strategies. *Proceedings of the National Academy of Sciences*, 105:10,057–60, http://www.pnas.org/content/105/29/10057.full; and Wason, E. L., and M. D. Hunter. 2014. Genetic variation in plant volatile emission does not result in differential attraction of natural enemies in the field. *Oecologia* 174:479–91.

Page 116: In the *Arthashastra*, written around the fourth century BCE, it was said that "a king whose territory has a common boundary with that of an antagonist is an ally." Some years later in a *Star Trek* film, Captain Kirk decided to join forces with an enemy in the battle against an even more fierce enemy, stating "the enemy of my enemy is friend." Spock was suspect of this approach and warned Kirk that this strategy, derived from an "Arab" proverb, was endorsed by a prince who was later decapitated by his so-called friend.

CHAPTER 6: SAVING UP TO RAISE A FAMILY

Page 119: Epigraph. John Burns, a well-known lepidopterist, wrote this poem in honor of Lincoln Brower et al.'s 1967 paper: Plant poisons in a terrestrial food chain. *Proceedings of the National Academy of Sciences* 57:893–98. The poem was published in his classic book of ecological poetry (1975), *BioGraffiti: A Natural Selection*. New York: Quadrangle/ New York Times Book Co.

Pages 121–122: Bernays studied the aposematic caterpillar of *Uresiphita reversalis* (Pyralidae); see Bernays, E. A. 1997. Feeding by lepidopteran larvae is dangerous. *Ecological Entomology* 22:121–23.

Pages 124–125: The history of ideas on sequestration and aposematic coloration are detailed in Slater, J. 1877. On the food of gaily-coloured caterpillars. *Transactions of the Royal Entomological Society of London* 25:205–9. Wallace also wrote about sequestration and aposematism: Wallace, A. R. 1889. *Darwinism*. London: Macmillan, https://www.gutenberg.org/ebooks/14558. It is perhaps remarkable that the codiscoverer of the theory of evolution by natural selection would thirty years later write a book called *Darwinism*. Wallace was a lifelong proponent of Darwin, and although he had slightly different views on certain aspects of evolution, he supported the general view of Darwin's *On the Origin of Species* as being one of the revolutionary advances in science. Finally, the English translation of Haase's book 1893 book appeared in 1896: Haase, E. *Researches on Mimicry*. Translated by C. M. Child. Stuttgart: Erwin Nägele; an electronic version, scanned from Cornell University's original copy can be found at http://books.google.com/books.

Pages 125–127: The few surveys of monarch mortality indicate variable, but very high levels of predation by various insect predators and parasitoids. Reviewed in Oberhauser, K. S., et al. 2015. Lacewings, wasps, and flies—Oh my, in K. Oberhauser, K. Nail, and S. Altizer, eds. *Monarchs in a Changing World: Biology and Conservation of an Iconic Insect*. Ithaca, NY: Cornell University Press, 71–82. A recent study discusses predators in monarch butterfly conservation: McCoshum, S. M., et al. Species distribution models for natural enemies of monarch butterfly (*Danaus plexippus*) larvae and pupae: Distribution patterns and implications for conservation. *Journal of Insect Conservation* 20:223–37.

Pages 125–127: Two studies, separated by twenty-five years, report that birds avoid monarchs and that aposematic coloration is the cause: Heinrich, B. (1979). Foraging strategies of caterpillars. *Oecologia* 42:325–37; and Hitchcock, C. B. 2004. Survival of caterpillars in the face of predation by birds: Predator-free space, caterpillar mimicry and protective coloration. PhD diss., University of Pennsylvania. More generally, a meta-analysis of fifty-one studies reports a large effect of bird removals on caterpillar abundance (including many species across many ecosystems): Mooney, K. A., et al. 2010. Interactions among predators and the cascading effects of vertebrate insectivores on arthropod communities and plants. *Proceedings of the National Academy of Sciences*, 107:7335–40, http://www.pnas.org/content/107/16/7335.full.

Pages 127–130: Selective sequestration of cardenolides by monarchs was observed and reported in several papers: Nelson, C. J. 1993. A model for cardenolide and cardenolide glycoside storage by the monarch butterfly, in S. B. Malcolm and M. P. Zalucki, eds. *Biology and Conservation of the Monarch Butterfly*. Los Angeles: Natural History Museum of Los Angeles County, 83–90. Taste perception of cardenolides was reported in Brower, L. P., and L. S. Fink. 1985. A natural toxic defense system: Cardenolides in butterflies versus birds. *Annals of the New York Academy of Sciences* 443:171–88. Invertebrates like praying mantises gut monarch caterpillars, as shown in Rafter, J. L., A. A. Agrawal, and E. L. Preisser. 2013. Chinese mantids gut toxic monarch caterpillars: Avoidance of prey defence? *Ecological Entomology* 38:76–82. Additionally, when assassin bug predators were force-fed monarchs that had been eating a highly toxic milkweed, the predators were eventually killed, but this took many days and eating many larvae: James, D. G. 2000. Feeding on larvae of *Danaus plexippus* (L.) (Lepidoptera: Nymphalidae) causes mortality in the assassin bug *Pristhesancus plagipennis* Walker (Hemiptera: Reduviidae). *Australian Entomologist* 27:5–8.

Page 129: Although first described in the 1950s, response of monarch caterpillars to sound was firmly established in Rothschild, M., and G. Bergström. 1997. The monarch butterfly caterpillar (*Danaus plexippus*) waves at passing hymenoptera and jet aircraft—Are repellent volatiles released simultaneously? *Phytochemistry* 45:1139–44. This phenomenon was studied in much greater detail a decade later in Taylor, C. J. 2008. Hearing in larvae of the monarch butterfly, *Danaus plexippus*, and selected other Lepidoptera. M.S. thesis, Carleton University.

Pages 130–132: The original notion of animal self-medication traces back to observations of my undergraduate mentor, Daniel Janzen, when he mused on why some mammals seemed to switch their diet to seemingly toxic foods: Janzen, D. H. 1978. Complications in interpreting the chemical defences of trees against tropical arboreal plant-eating vertebrates, in G. G. Montgomery, ed. *The Ecology of Arboreal Folivores*. Washington, DC: Smithsonian Institution Press, 73–84. Since then, the term "zoopharmacognosy" (read as "animal" "drug" "knowing") has been applied to the study of nonhuman self-medication. The work on monarchs (and the basis for fig. 6.3) are taken from work by Jaap de Roode's lab at Emory University. Two important studies on self-medication in monarchs include Lefevre, T., et al. 2010. Evidence for transgenerational medication in nature. *Ecology Letters* 13:1485–93; and Lefevre, T., et al. 2012. Behavioural resistance against a protozoan parasite in the monarch butterfly. *Journal of Animal Ecology* 81:70–79, http://onlinelibrary.wiley.com/doi/10.1111/j.1365-2656.2011.01901.x/full.

Page 141: Viceroy mimicry was reported in van Zandt Brower, J. 1958. Experimental studies of mimicry in some North American butterflies: Part I. The monarch, *Danaus plexippus*, and viceroy, *Limenitis archippus archippus*. *Evolution:* 32–47. This was extended by Ritland, D. B., and L.P. Brower. 2000. Mimicry-related variation in wing color of viceroy butterflies (*Limenitis archippus*): A test of the model-switching hypothesis (Lepidoptera: Nymphalidae). *Holarctic Lepidoptera* 7:5–11; and Ritland, D. B. 1998. Mimicry-related predation on two viceroy butterfly (*Limenitis archippus*) phenotypes. *American Midland Naturalist* 140:1–20.

Pages 142–143: Where cardenolides are stored in monarchs was summarized early on by Brower, L. P., and S. C. Glazier. 1975. Localization of heart poisons in the monarch butterfly. *Science* 188:19–25. This study confirmed the early work of Parsons (see chapter 2) showing that cardenolides are concentrated in adult wings of monarchs. This work was further elaborated in Nishio, S. 1980. The fates and adaptive significance of cardenolides sequestered by larvae of *Danaus plexippus* (L.) and *Cycnia inopinatus* (Hy. Edwards). PhD diss., University of Georgia.

CHAPTER 7: THE MILKWEED VILLAGE

Page 148: Epigraph. Muir, John. 1911. *My First Summer in the Sierra*. Boston: Houghton Mifflin.

Pages 149–150: Recent work on the evolutionary history of insects and the earliest known herbivory are in Misof, B., et al. 2014. Phylogenomics resolves the timing and pattern of insect evolution. *Science* 346:763–67; and Labandeira, C. 2007. The origin of herbivory on land: Initial patterns of plant tissue consumption by arthropods. *Insect Science* 14:259–75, http://onlinelibrary.wiley.com/doi/10.1111/j.1744-7917.2007.00141.x-i1/epdf.

Pages 150–152: The goldenrod insect fauna was summarized in Root, R. B. 1996. Herbivore pressure on goldenrods (*Solidago altissima*): Its variation and cumulative effects. *Ecology* 77:1074–87. Root defined this community of more than a hundred species as completing a substantial part of their development eating tall goldenrod. Root's last conceptualization of the "guild" can be found in Root, R. B. 2001. Guilds, in S. Levin, ed. *Encyclopedia of Biodiversity*, vol. 3. San Diego, CA: Academic Press, 295–302. The quotation in the text is from this article.

Pages 153–155: An undergraduate student in my laboratory, Alex Smith, studied the differences and coexistence of the three milkweed species; his research was summarized in Smith, R. A., K. A. Mooney, and A. A. Agrawal. 2008. Coexistence of three specialist aphids on common milkweed, *Asclepias syriaca*. *Ecology* 89:2187–96. We explored the role of plant traits in defending against *Aphis asclepiadis* by altering ant tending (as well as the indirect effects of ant tending on monarch survival) in Mooney, K. A., and A. A. Agrawal. 2008. Plant genotype shapes ant-aphid interactions: Implications for community structure and indirect plant defense. *American Naturalist* 171:E195–E205, http://www.jstor.org/stable/10.1086/587758. More recent exploration of processing and sequestration of cardenolides by milkweed aphids may be found in Züst, T., and A. A. Agrawal. 2015. Population growth and sequestration of plant toxins along a gradient of specialization in four aphid species on the common milkweed *Asclepias syriaca*. *Functional Ecology* 30:547–56.

Pages 156–158: The most recent information we have on the expanse of *Tetraopes* beetles is provided in Farrell, B. D. 2001. Evolutionary assembly of the milkweed fauna: Cytochrome oxidase I and the age of *Tetraopes* beetles. *Molecular Phylogenetics and Evolution* 18:467–78. Current work in Brian Farrell's lab holds promise to revise and update the evolutionary relationships among *Tetraopes* species. Much is known about their ecology, however, including the following two publications: Agrawal, A. A. 2004. Resistance and susceptibility of milkweed: Competition, root herbivory, and plant genetic variation. *Ecology* 85:2118–33; and Rasmann, S., et al. 2011. Direct and indirect root defense of milkweed (*Asclepias syriaca*): Trophic cascades, tradeoffs, and novel methods for studying subterranean herbivory. *Journal of Ecology* 99:16–25, http://onlinelibrary.wiley.com/doi/10.1111/j.1365-2745.2010.01713.x/full.

Pages 161–162: The classic study on deactivation of milkweed latex is Dussourd, D. E., and T. Eisner. 1987. Vein-cutting behavior: Insect counterploy to the latex defense of plants. *Science* 237:898–901.

Pages 163–164: Our work on the convergence of molecular substitutions in the milkweed insects was published in Dobler, S., et al. 2012. Community-wide convergent evolution in insect adaptation to toxic cardenolides by substitutions in the Na, K-ATPase. *Proceedings of the National Academy of Sciences* 109:13,040–45, http://www.pnas.org/content/109/32/13040.full. A group working independently confirmed our results, published two months later: Zhen, Y., et al. 2012. Parallel molecular evolution in an herbivore community. *Science* 337:1634–37.

Pages 166–167: For each entry in *Species Plantarum* (http://biodiversitylibrary.org/page/358012), Linnaeus listed past scientific names as a way of synonymizing names and connecting the literature. For *Asclepias syriaca* (p. 214), he incorrectly listed a past name as "Apocynum Syriacum. Clus. hist. 2. p. 87," referring to Carolus Clusius's 1601 description of the Middle Eastern species Beidelsar. Cornuti's mistake, confusing Beidelsar/*Apocynum syriacum* and common milkweed, was thus propagated by Linnaeus. Studies of insects on Beidelsar, now known as *Calotropis*, are summarized in Dhileepan, K. 2014. Prospects for the classical biological control of *Calotropis procera* (Apocynaceae) using co-evolved insects. *Biocontrol Science and Technology* 24: 977–98.

Page 168: A milkweed-feeding grasshopper was originally studied by Reichstein and Rothschild, see Euw, J. V., et al. 1967. Cardenolides (heart poisons) in a grasshopper feeding on milkweeds. *Nature* 214:35–39. Although the molecular basis of cardenolide resistance is unknown, physiologically it has been shown that the grasshopper *Poekilocerus bufonius* has a relatively insensitive sodium pump: see Al-Robai, A. A. 1993. Different ouabain sensitivities of Na^+/K^+-ATPase from *Poekilocerus bufonius* tissues and a possible physiological cost. *Comparative Biochemistry and Physiology Part B: Comparative Biochemistry* 106:805–12. The cardenolide-insensitive wasp was documented in Dobler, S., et al. 2015. Convergent adaptive evolution—How insects master the challenge of cardiac glycoside-containing host plants. *Entomologia Experimentalis et Applicata* 157:30–39, http://onlinelibrary.wiley.com/doi/10.1111/eea.12340/full.

Page 168: The phrase "entangled bank" was introduced by Darwin in his 1859 book, *On the Origin of Species*. He began the last paragraph of the book: "It is interesting to contemplate an entangled bank, clothed with many plants of many kinds, with birds singing on the bushes, with various insects flitting about, and with worms crawling through the damp earth, and to reflect that these elaborately constructed forms, so different from each other, and dependent on each other in so complex a manner, have all been produced by laws acting around us." By the fifth edition of the book (published in 1869) "entangled bank" was replaced with "tangled bank." The complete works of Darwin are available at http://darwin-online.org.uk.

Pages 174–176: Various aspects of human use of milkweed have been detailed in the following: Gaertner, E. E. 1979. The history and use of milkweed (*Asclepias syriaca* L.). *Economic Botany*, 33:119–23; Small, E. 2015. Milkweeds—A sustainable resource for humans and butterflies. *Biodiversity* 16:290–303; and Schwartz, D. M. 1987. Underachiever of the plant world. *Audubon* 89:46–61 (from which the quotation about fabrics is taken).

Page 176: On the importance of cooking in human evolution, see Wrangham, R. 2009. *Catching Fire: How Cooking Made Us Human*. New York: Basic Books. Euell Gibbons's article, How to milk a milkweed, was published in *Organic Gardening*, January 1972, 148–53.

CHAPTER 8: THE AUTUMN MIGRATION

Page 178: Epigraph. Valerie Dohren, "The Migration." Used by permission of the author.

Pages 180–181: Work on the role of juvenile hormone and its impacts on monarch reproductive diapause, longevity, and fat storage is summarized in Herman W. S., and M. Tatar. 2001. Juvenile hormone regulation of longevity in the migratory monarch butterfly. *Proceedings of the Royal Society B* 268:2509–14. The role of day length, temperature, and plant quality on reproductive diapause is summarized in: Goehring, L. and K. S. Oberhauser. 2002. Effects of photoperiod, temperature, and host plant age on induction of reproductive diapause and development time in *Danaus plexippus*. *Ecological Entomology* 27:674–85.

Pages 181–182: The southern flight in monarchs is apparently unrelated to the physiological changes induced by the lack of juvenile hormone: see Zhu, H., et al. 2009. Defining behavioral and molecular differences between summer and migratory monarch butterflies. *BMC Biology* 7:14, http://bmcbiol.biomedcentral.com/articles/10.1186/1741-7007-7-14.

Pages 182–183: Work on the shape and size of migratory and nonmigratory monarchs is published in Altizer, S., and A. K. Davis. 2010. Populations of monarch butterflies with different migratory behaviors show divergence in wing morphology. *Evolution* 64:

1018–28. David Gibo summarized his estimates of monarch flying height in Gibo, D. L. 1981. Altitudes attained by migrating monarch butterflies, *Danaus plexippus* (Lepidoptera: Danaidae), as reported by glider pilots. *Canadian Journal of Zoology* 59:571–72. Other work on the energetics of monarch flight and the benefits of soaring and gliding were published in Gibo, D. L., and M. J. Pallett. 1979. Soaring flight of monarch butterflies, *Danaus plexippus* (Lepidoptera: Danaidae), during the late summer migration in southern Ontario. *Canadian Journal of Zoology* 57:1393–1401.

Pages 185–189: Our current understanding on how monarchs navigate their southern migration was reviewed in Reppert, S. M., P. A. Guerra, and C. Merlin. 2016. Neurobiology of monarch butterfly migration. *Annual Review of Entomology* 61:25–42. The original work on the antennal clock was reported in Merlin, C., R. J. Gegear, and S. M. Reppert. 2009. Antennal circadian clocks coordinate sun compass orientation in migratory monarch butterflies. *Science* 325:1700–1704. Classic work on the sun compass from birds and bees was published more than fifty years ago: Kramer, G. 1952. Experiments on bird orientation. *Ibis* 94:265–85; and Lindauer, M. 1960. Time-compensated sun orientation in bees. *Cold Spring Harbor Symposia on Quantitative Biology* 25:371–77. The first study demonstrating that monarchs use a time-compensated sun compass was Perez, S. M., O. R. Taylor, and R. Jander. 1997. A sun compass in monarch butterflies. *Nature* 387:29. In this very brief report, many details of study were not included, and the data have been questioned. Nonetheless, the study has stood the test of time, and more rigorous, controlled, mechanistic studies definitively revealed that monarchs do, indeed, use the time-compensated sun-compass: Mouritsen, H., and B. G. Frost. 2002. Virtual migration in tethered flying monarch butterflies reveals their orientation mechanisms. *Proceedings of the National Academy of Sciences*, 99:10,162–66, http://www.pnas.org/content/99/15/10162.full.

Pages 189–192: The issues of magnetic maps as well as details of experiments with sea turtles and birds are in Lohmann, K. J., C. M. Lohmann, and N. F. Putman. 2007. Magnetic maps in animals: Nature's GPS. *Journal of Experimental Biology* 210:3697–705, http://jeb.biologists.org/content/jexbio/210/21/3697.full.pdf. Work with monarchs is summarized in several papers: Calvert, W. H. 2001. Monarch butterfly (*Danaus plexippus* L., Nymphalidae) fall migration: Flight behavior and direction in relation to celestial and physiographic cues. *Journal of the Lepidopterists' Society* 55:162–68, http://images.peabody.yale.edu/lepsoc/jls/2000s/2001/2001-55(4)162-Calvert.pdf; Guerra, P. A., R. J. Gegear, and S. M. Reppert. 2014. A magnetic compass aids monarch butterfly migration. *Nature Communications* 5, article no. 4164, http://www.nature.com/ncomms/2014/140624/ncomms5164/full/ncomms5164.html; Rogg, K. A., O. R. Taylor, and D. L. Gibo. 1999. Mark and recapture during the monarch migration: A preliminary analysis, in W. A. Haber et al., eds. *1997 North American Conference on the Monarch Butterfly*. Montreal: Commission for Environmental Cooperation, 133–38; and for figure 8.5, Urquhart, F. A. 1964/1965. Monarch butterfly (*Danaus plexippus*) migration studies: Autumnal movement. *Proceedings of the Entomological Society of Ontario* 95:23–33.

Page 197: Pyle quoted in Blakeslee, S. 1986. Butterfly seen in new light by scientists. *New York Times*, November 28.

Pages 197–199: Brower reviewed the history of the monarch migration, including his own experiences, in Brower, L. P. 1995. Understanding and misunderstanding the migration of the monarch butterfly (Nymphalidae) in North America: 1857–1995. *Journal of the Lepidopterists' Society* 49:304–85, http://images.peabody.yale.edu/lepsoc/jls/1990s/1995/1995-49(4)304-Brower.pdf.

Pages 200–201: A summary of the monarch overwintering sites and their geography is in Slayback, D. A., et al. 2007. Establishing the presence and absence of overwintering colonies of the monarch butterfly in Mexico by the use of small aircraft. *American Entomologist* 53:28–41.

Page 204: A general summary of the intersection of migration and disease is provided in Altizer, S., R. Bartel, and B. A. Han. 2011. Animal migration and infectious disease risk. *Science* 331:296–302. Recent work on parasite loads in migratory and nonmigratory monarch populations is summarized in Satterfield, D. A., J. C. Maerz, and S. Altizer. 2015. Loss of migratory behaviour increases infection risk for a butterfly host. *Proceedings of the Royal Society of London B: Biological Sciences*, 282:1734, http://rspb.royalsocietypublishing.org/node/65352.full. Finally, work showing an important role for the migration in monarch-parasite interaction is Bartel, R., et al. 2011. Monarch butterfly migration and parasite transmission in eastern North America. *Ecology* 92:342–51.

Page 205: Work on monarch thermoregulation and impacts on lipids was published in Masters, A. R., S. B. Malcolm, and L. P. Brower. 1988. Monarch butterfly (*Danaus plexippus*) thermoregulatory behavior and adaptations for overwintering in Mexico. *Ecology* 458–67.

Pages 205–208: Work on bird predation on monarchs at the overwintering sites was presented in two classic studies: Calvert, W. H., L. E. Hedrick, and L. P. Brower. 1979. Mortality of the monarch butterfly (*Danaus plexippus* L.): Avian predation at five overwintering sites in Mexico. *Science* 204:847–51; and Fink, L. S., and L. P. Brower. 1981. Birds can overcome the cardenolide defence of monarch butterflies in Mexico. *Nature* 291:7–13. Studies on mice predation of monarchs was summarized in Glendinning, J. I. 1993. Comparative feeding responses of the mice *Peromyscus melanotis*, *P. aztecus*, *Reithrodontomys sumichrasti*, and *Microtus mexicanus* to overwintering monarch butterflies in Mexico, in S. B. Malcolm and M. Zalucki, eds. *Biology and Conservation of the Monarch Butterfly*. Los Angeles: Natural History Museum of Los Angeles County, 323–33.

CHAPTER 9: LONG LIVE THE MONARCHY!

Page 210: Epigraph. Aridjis, Homero. 2002. "About Angels IX," trans. George McWhirter, in George McWhirter and Betty Ferber, eds. *Eyes to See Otherwise*. New York: New Directions; and in *A Time of Angels*. San Francisco: City Lights, 2012.

Pages 211–213: There have long been claims about declines of monarch populations, but the first strong and statistically analyzed data were presented in Brower, L. P., et al. 2012. Decline of monarch butterflies overwintering in Mexico: Is the migratory phenomenon at risk? *Insect Conservation and Diversity* 5:95–100, http://onlinelibrary.wiley.com/doi/10.1111/j.1752-4598.2011.00142.x/full. In the same year, an alternative analysis and conclusion was published using different data: Davis, A. K. 2012. Are migratory monarchs really declining in eastern North America? Examining evidence from two autumn census programs. *Insect Conservation and Diversity* 5:101–5, http://onlinelibrary.wiley.com/doi/10.1111/j.1752-4598.2011.00158.x/full. A recent analysis of the western population of monarchs shows a forty-year decline in this population as well: Espeset, A., et al. 2016. Understanding a migratory species in a changing world: Climatic effects and demographic declines in the western monarch revealed by four decades of intensive monitoring. *Oecologia* 181:819–30. Finally, many of the world's great mammal and bird migrations are threatened; see the following two reviews: Bowlin, M.S., et al. 2010. Grand challenges in migration biology. *Integrative and Comparative Biology* 50:261–79, http://icb.oxfordjournals.org/content/50/3/261.full.pdf+html; and Wilcove, D.S., and M. Wikelski. 2008. Going, going, gone: Is animal migration disappearing? *PLoS Biology* 6:e188, http://dx.doi.org/10.1371/journal.pbio.0060188.

Page 216: The first study from our work on the social aspects of the monarch was published in Gustafsson, K. M., et al. 2015. The monarch butterfly through time and space: The social construction of an icon. *BioScience* 65:112–22, http://bioscience.oxfordjournals.org/content/early/2015/05/04/biosci.biv045.full.

Page 217: An excellent study documenting the impacts of spring climate on the summer breeding populations of monarchs is Zipkin, E. F., et al. 2012. Tracking climate impacts on the migratory monarch butterfly. *Global Change Biology* 18:3039–49. Our own analysis of monarch population dynamics is published in Inamine, H., et al. Linking the continental migratory cycle of the monarch butterfly to understand its population decline. *Oikos* 125:1081–1091. http://onlinelibrary.wiley.com/doi/10.1111/oik.03196/full.

Page 219: Excellent summaries of logging at the monarchs' overwintering sites and changes in forest cover are provided in the following: Keiman, A. F., and M. Franco. 2004. Can't see the forest for the butterflies: The need for understanding forest dynamics at the monarch's overwintering sites, in K. S. Oberhauser and M. J. Solensky, eds. *Monarch Butterfly Biology and Conservation*. Ithaca, NY: Cornell University Press, 135–40; and Shahani, P. C., et al. 2015. Monarch habitat conservation across North America, in K. S. Oberhauser, K. R. Nail, and S. Altizer, eds. *Monarchs in a Changing World: Biology and Conservation of an Iconic Butterfly*. Ithaca, NY: Cornell University Press, 31–41. On forest history, see Brower, L. P., et al. 2002. Quantitative changes in forest quality in a principal overwintering area of the monarch butterfly in Mexico, 1971–1999. *Conservation Biology* 16:346–59; and Vidal, O., J. López-García, and E.

Rendón-Salinas. 2014. Trends in deforestation and forest degradation after a decade of monitoring in the Monarch Butterfly Biosphere Reserve in Mexico. *Conservation Biology*, 28:177–86, http://onlinelibrary.wiley.com/doi/10.1111/cobi.12138/full.

Pages 219–220: The role of trees as jackets, umbrellas, and hot water bottles is in Williams, E. H., and L. P. Brower. 2015. Microclimatic protection of overwintering monarchs provided by Mexico's high-elevation Oyamel fir forests: A review, in Oberhauser, Nail, and Altizer. *Monarchs in a Changing World*, 109–16.

Page 221: Lopez-Portillo, J. L. 1980. Presidential decree protecting the monarch in all parts of Mexico. *Diario Oficial* (Mexico City), March 25, 106–7.

Page 221: International Union for Conservation of Nature and Natural Resources. 1983. *The IUCN Invertebrate Red Data Book*. Gland, Switzerland: IUCN, 463–70.

Pages 221–223: Vidal, López-García, and Rendón-Salinas. 2014. Trends in deforestation and forest degradation after a decade of monitoring in the Monarch Butterfly Biosphere Reserve in Mexico. *Conservation Biology*, 28:177–86.

Page 224: The impacts of climate change on Oyamel fir tree distributions and potential effects on monarchs is summarized in two publications: Oberhauser, K., and A. T. Peterson. 2003. Modeling current and future potential wintering distributions of eastern North American monarch butterflies. *Proceedings of the National Academy of Sciences* 100:14,063–68, http://www.pnas.org/content/100/24/14063.full; and Sáenz-Romero, C., et al. 2012. *Abies religiosa* habitat prediction in climatic change scenarios and implications for monarch butterfly conservation in Mexico. *Forest Ecology and Management* 275:98–106.

Pages 225–226: Taylor quoted at Monarch Watch Update, February 21, 2003, http://www.monarchwatch.org.

Page 227: Maeckle, M. 2014. First Lady Michelle Obama gets milkweed as White House adds first pollinator garden. Texas Butterfly Ranch Blog post, April 6, http://texasbutterflyranch.com/2014/04/06/first-lady-michelle-obama-gets-milkweed-as-white-house-adds-first-pollinator-garden.

Page 227: Urquhart, F., and N. Urquhart. 1979. *Annual Insect Migration Studies Newsletter*, vol. 16, http://www.monarchwatch.org/read/articles/index.htm; Blakeslee, S. 1986. Butterfly seen in new light by scientists. *New York Times*, November 28; and Yoon, C. K. 1998. On the trail of the monarch, with the aid of chemistry. *New York Times*, December 29.

Page 233: National Academies of Sciences, Engineering, Medicine. 2016. *Genetically Engineered Crops: Experiences and Prospects*. Washington, DC: National Academies Press, 95–96, https://nas-sites.org/ge-crops/2016/05/17/report.

Page 238: Sex ratio changes in monarchs at the overwintering sites are in Davis, A. K., and E. Rendón-Salinas. 2009. Are female monarch butterflies declining in eastern North America? Evidence of a 30-year change in sex ratios at Mexican overwintering sites. *Biology Letters*, rsbl20090632, http://rsbl.royalsocietypublishing.org/content/early/2009/09/17/rsbl.2009.0632.short.

Page 239: A summary of drought effects on monarchs is in Brower, L. P., et al. 2015. Effect of the 2010–2011 drought on the lipid content of monarchs migrating through Texas to overwintering sites in Mexico, in Oberhauser, Nail, and Altizer. *Monarchs in a Changing World*, 117–29. Further modeling work indicating the importance of spring precipitation is in Zipkin et al. 2012. Tracking climate impacts on the migratory monarch butterfly. *Global Change Biology* 18:3039–49.

Page 240: Other threats to monarchs have been widely discussed. Ecotourism in particular is mentioned in Vidal, O., and E. Rendón-Salinas. (2014). Dynamics and trends of overwintering colonies of the monarch butterfly in Mexico. *Biological Conservation* 180:165–75. Preliminary studies of neonicitinoid insecticides on monarchs are reported in Krischik, V. A., et al. 2015. Soil-applied imidacloprid is translocated to ornamental flowers and reduces survival of adult *Coleomegilla maculata*, *Harmonia axyridis*, and *Hippodamia convergens* lady beetles, and larval *Danaus plexippus* and *Vanessa cardui*. *PLoS One*, doi: 10.1371/journal.pone.0119133. The issue of automobile accidents is studied in McKenna, D. D., et al. 2001. Mortality of Lepidoptera along roadways in central Illinois. *Journal of the Lepidopterists' Society*, 55:63–68, http://images.peabody.yale.edu/lepsoc/jls/2000s/2001/2001-55(2)63-McKenna.pdf. May Berenbaum revisited this issue in an essay after the petition to list monarchs as threatened under the endangered species act: Berenbaum, M. 2015. Road worrier. *American Entomologist* 61:5–8.

IMAGE CREDITS

Fig. 1.1. Photos by Ellen Woods.
Fig. 1.2. Photo by Anurag Agrawal.
Fig. 1.3. Original artwork by Frances Fawcett.
Fig. 1.4. Based on an image produced by ProQuest LLC as part of ProQuest® Historical Newspapers (http://www.proquest.com) and reproduced with permission. Further reproduction is prohibited without permission.
Fig. 1.5. Photo by Daniel H. Janzen, from the course he teaches; see http://fission.sas.upenn.edu/caterpillar.
Fig. 1.6. Original artwork by Frances Fawcett, based on Wheeler, Q. D. 1990. Insect diversity and cladistic constraints. *Annals of the Entomological Society of America* 83: 1031–47.
Fig. 1.7. Photo by Ellen Woods.
Fig. 1.8. Image by Georg Petschenka.
Fig. 1.9. Photos by Anurag Agrawal, except the photo of *Asclepias lemmonii* (by Mark Fishbein).
Fig. 2.1. Photo by Anurag Agrawal.
Fig. 2.2. Original artwork by Frances Fawcett.
Fig. 2.3. Photos (a) and (c) by Ellen Woods. Photo (b) by Dwight Kuhn. Photo (d) by Kevin Berwin.
Fig. 2.4. Based on an image produced by ProQuest LLC as part of ProQuest® Historical Newspapers (http://www.proquest.com) and reproduced with permission. Further reproduction is prohibited without permission.
Fig. 2.5. Photos by Lincoln P. Brower, Sweet Briar College.
Fig. 2.6. Original artwork by Frances Fawcett, based on fig. 5.3 in Karban, R., and I. T. Baldwin. 1997. *Induced responses to herbivory*. Chicago: University of Chicago Press.
Fig. 2.7. Image by Frances Fawcett, based on an imaged provided by Tobias Züst.
Fig. 3.1. Reproduced with permission from the American Medical Association, the January 20, 1981, cover image of *JAMA*, copyright American Medical Association.
Fig. 3.2. Photo (a) by Anurag Agrawal; photo (b) by Joseph Haz Hall.
Fig. 3.3. Images by Meena Haribal and Georg Petschenka.
Fig. 3.4. Original artwork by Frances Fawcett.
Fig. 3.5. Photo by Anurag Agrawal.
Fig. 3.6. Photos by Georg Petschenka, modified by Frances Fawcett.
Fig. 3.7. Image by Georg Petschenka, except photo of garden tiger moth (by Robert Trusch).
Fig. 4.1. Original artwork by Frances Fawcett.
Fig. 4.2. Sky photo by Richard Ellis; stream photo by Anurag Agrawal.

Fig. 4.3. Artwork by Frances Fawcett, based on Guerra, P. A., and S. M. Reppert. 2013. Coldness triggers northward flight in remigrant monarch butterflies. *Current Biology* 23:419–23.

Fig. 4.4. Original artwork by Frances Fawcett.

Fig. 4.5. Original artwork by Frances Fawcett, based on Oberhauser, K., and D. Frey. 1999. Coercive mating by overwintering male monarch butterflies, in W. A. Haber et al., eds. *1997 North American Conference on the Monarch Butterfly*. Montreal: Commission for Environmental Cooperation, 67–78.

Fig. 4.6. Original artwork by Frances Fawcett, based on figs. 8 and 10 in Urquhart, F. A. 1955. *Report on the studies of the movements of the monarch butterfly in North America*. Toronto: Royal Ontario Museum.

Fig. 4.7. Artwork by Frances Fawcett, based on Flockhart, D. T., et al. 2013. Tracking multi-generational colonization of the breeding grounds by monarch butterflies in eastern North America. *Proceedings of the Royal Society B* 280:1087.

Fig. 4.8. Original artwork by Frances Fawcett; photos by Anurag Agrawal.

Fig. 4.9. Photo by Ellen Woods.

Fig. 4.10. Original artwork by Frances Fawcett.

Fig. 5.1. Photo (a) by Kailen Mooney; photo (b) by Ellen Woods; photo (c) by Anurag Agrawal.

Fig. 5.2. Photo (a) by Ellen Woods; photo (b) by Anurag Agrawal and Carole S. Daugherty.

Fig. 5.3. Photo by Ellen Woods and graph by Frances Fawcett, based on data from Zalucki, M. P., and S. B. Malcolm. 1999. Plant latex and first-instar monarch larval growth and survival on three North American milkweed species. *Journal of Chemical Ecology* 25:1827–42.

Fig. 5.4. Photos by Ellen Woods.

Fig. 5.5. Photo (a) by Anurag Agrawal; photo (b) by Ellen Woods.

Fig. 5.6. Original artwork by Frances Fawcett, based on Zalucki, M. P., and L. P. Brower. 1992. Survival of first instar larvae of *Danaus plexippus* (Lepidoptera: Danainae) in relation to cardiac glycoside and latex content of *Asclepias humistrata* (Asclepiadaceae). *Chemoecology* 3:81–93.

Fig. 5.7. Photos by Ellen Woods.

Fig. 5.8. Graphs drawn by Frances Fawcett, based on the author's research.

Fig. 5.9. Graphs drawn by Frances Fawcett, based on the author's research.

Fig. 5.10. Graphs drawn by Frances Fawcett, based on the author's research.

Fig. 6.1. Photo (a) by Ellen Woods; photo (b) by Anurag Agrawal.

Fig. 6.2. Photo (a) by Ellen Woods; photo (b) by Anurag Agrawal; photo (c) by Jaap de Roode.

Fig. 6.3. Artwork by Frances Fawcett, based on Lefevre, T., et al. 2012. Behavioural resistance against a protozoan parasite in the monarch butterfly. *Journal of Animal Ecology* 81:70–79.

Fig. 6.4. Photos by Ellen Woods.
Fig. 6.5. Photos by Ellen Woods.
Fig. 6.6. Photos by Ellen Woods.
Fig. 6.7. Photos (a) and (c) by Ellen Woods; photos (b) and (d) by Anurag Agrawal.
Fig. 6.8. Artwork by Frances Fawcett, based on data from Petschenka, G., and A. A. Agrawal. 2015. Milkweed butterfly resistance to plant toxins is linked to sequestration, not coping with a toxic diet. *Proceedings of the Royal Society B* 282:1865. Caterpillar images by Georg Petschenka.
Fig. 7.1. Photos by Ellen Woods.
Fig. 7.2. Photos (a) and (c) by Ellen Woods; photo (b) by Anurag Agrawal.
Fig. 7.3. Photo by Ellen Woods.
Fig. 7.4. Photos (a), (b), (c) by Anurag Agrawal; photo (d) by Ellen Woods.
Fig. 7.5. Original artwork by Frances Fawcett.
Fig. 7.6. Photo by Anurag Agrawal.
Fig. 7.7. Original artwork by Frances Fawcett.
Fig. 7.8. Photo by Anurag Agrawal.
Fig. 7.9. Photo by Anurag Agrawal.
Fig. 8.1. Photo by Kent Weakley.
Fig. 8.2. Original artwork by Frances Fawcett, based on Gibo, D. L., and M. J. Pallett. 1979. Soaring flight of monarch butterflies, *Danaus plexippus* (Lepidoptera: Danaidae), during the late summer migration in southern Ontario. *Canadian Journal of Zoology* 57:1393–1401.
Fig. 8.3. Photos by Jim Ellis.
Fig. 8.4. Artwork by Frances Fawcett, based on Reppert, S. M. 2007. The ancestral circadian clock of monarch butterflies: Role in time-compensated sun compass orientation. *Cold Spring Harbor Symposia on Quantitative Biology* 72:113–18.
Fig. 8.5. Artwork by Frances Fawcett, based on Urquhart, F. A. 1964/1965. *Proceedings of the Entomological Society of Ontario* 95:23–33.
Fig. 8.6. Image by Albert Moldvay/National Geographic Creative.
Fig. 8.7. Photos by Kent Weakley.
Fig. 8.8. Photo by Anurag Agrawal.
Fig. 8.9. Photo by Lee Hedrick, owned and provided by Lincoln P. Brower, Sweet Briar College.
Fig. 8.10. Original artwork by Jay Hart (http://earthpattern.com).
Fig. 8.11. Photos by Lincoln P. Brower, Sweet Briar College.
Fig. 8.12. Original artwork by Frances Fawcett, based on the cover image of Fink, L. S., and L. P. Brower. 1981. Birds can overcome the cardenolide defence of monarch butterflies in Mexico. *Nature* 291:7–13.
Fig. 9.1. Graph drawn by Frances Fawcett, based on the author's research.
Fig. 9.2. Graphs drawn by Frances Fawcett, based on the author's research.
Fig. 9.3. Photo by Lincoln P. Brower, Sweet Briar College.

Fig. 9.4. Photo by Ellen Sharp.

Fig. 9.5. Artwork by Frances Fawcett, based on Vidal, O., J. López-García, and E. Rendón-Salinas. 2014. Trends in deforestation and forest degradation after a decade of monitoring in the Monarch Butterfly Biosphere Reserve in Mexico. *Conservation Biology* 28: 177–86.

Fig. 9.6. Photo by Lincoln P. Brower, Sweet Briar College.

Fig. 9.7. Original artwork by Frances Fawcett.

Fig. 9.8. Graphs drawn by Frances Fawcett, based on the author's research.

Acknowledgments images: Photos by Anurag Agrawal.

INDEX

Page numbers in *italics* refer to illustrations